Between One Culture

Robert Schiller

Between One Culture

Essays on Science and Art

 Springer

Robert Schiller
Centre for Energy Research, Hungarian Academy of Sciences
Budapest, Hungary

Language Editor: Bálint Gárdos

Selection from the volumes by the author in Hungarian
Egy kultúra között [Between one culture] 2004 and
A kételkedés gyönyörűsége [The pleasure of doubt] 2017
Typotex, Budapest

ISBN 978-3-030-20540-9   ISBN 978-3-030-20538-6   (eBook)
https://doi.org/10.1007/978-3-030-20538-6

Fundamentally revised from original Hungarian editions published by Typotex, Budapest, 2004 and 2017

This Springer imprint is published by the registered company Springer Nature Switzerland AG.
The registered company address is: Gewerbestrasse 11, 6330 Cham, Switzerland

*…those who assert that the mathematical sciences
say nothing of the beautiful or the good are in error.*
Aristotle, Metaphysics Book XIII

To Vera

Preface

We have one skull per neck and our intellectual husbandry must rest upon this stock. All the tasks where the brain is needed must be solved using this single tool. Thus, it looks somewhat wasteful to make use of our brain for only half of the problems we might encounter: to maintain that we are *either* science-minded *or* art-minded people. A part of the writings in this volume debates this dichotomy by finding and enjoying cases where these two great realms of intellectual activity are intertwined in one way or another. It is, however, not my aim to bridge the gap between the "two cultures". I am convinced that there exists only one single culture.

Being a chemist by training and profession and now writing about questions which are often far from that field, I do know that the results might inevitably be amateurish. In the days of professionalism, it is shameful to be a dilettante. Nevertheless, let us recall that this word of abuse is connected to a Latin verb meaning to love something, to esteem it or to hold it dear.

Plato thought that human soul had a rational, calculating part and a desiring, irrational part. He, however, who constructed a rational picture of the motion of the heavenly bodies also composed the lines:

Thou gazest on the stars, my star!
Ah! would that I might be
Myself those skies with myriad eyes,
That I might gaze on thee.

The two parts of Plato's soul kept close friendship.

Budapest, Hungary   Robert Schiller

Contents

About the Author

Robert Schiller (1935), born in Budapest, Hungary, graduated from R. Eötvös University, Budapest, in 1958 and completed his PhD in 1966 and DSc in 1974. He is a titular professor at R. Eötvös University and Dr. habil Privatdozent at Budapest Technical University. After completing his studies, Professor Schiller joined the Chemistry Department of the Central Research Institute for Physics, Hungarian Academy of Sciences, where he is now a Research Professor Emeritus. Having worked at several laboratories abroad, he e.g. spent a full year at the Paterson Laboratories, Manchester, UK. His main research interests are in radiation chemistry, electrochemistry and the theory of transport processes. Currently, he is investigating the effects of fast ions on metals. He has taught courses on radiation chemistry and statistical mechanics at R. Eötvös University and has published several textbooks in these areas. Apart from his research papers, Professor Schiller has also written books and a number of essays popularizing science. He was awarded the Wigner Prize by the Hungarian Academy of Sciences in 2001 and was voted the popular science author for the year 2012, and asteroid no.196005 was named Robertschiller in his honour.

Part I

Mythology

1

Is the Sea Wine-dark?

But the sea is blue! Wine-dark, this is an expression of Homer's, appearing quite often both in the *Iliad* and the *Odyssey*. This is not the only epithet connected to the sea and this epithet is also used in another context: oxen are wine-dark as well. The exact meaning of the Greek expression has often been debated, in different languages and different translations it appears as wine-coloured, wine-red, of wine-like surface or even bleak. Whatever solution the translator chooses, the word never indicates anything blue, in contrast to the readers' habitual expectation.

In the second part of the nineteenth century, when scientific rationalism held sway over the general ways of thinking, the epithet was explained in terms of natural history, psychology, linguistics and other sciences. Being related to seeing the obvious first idea was to think of the tradition of Homer's blindness. The much-quoted lines of the "Hymn to Apollo" may refer to the poet himself: "*It is a blind man dwelling in Chios, rugged and rocky, whose songs, every one, are the best both now and hereafter.*" (Translated by Rodney Merrill.) Still, it seems impossible to think that he who created the most gorgeous panorama of the ancient Mediterranean was a blind man.

I think that Gladstone, the future prime minister of England, was the first to suggest that perhaps not only the poet but the archaic Hellas as a whole was colour-blind. That is, colour vision, at least in the complete spectrum visible for present-day human beings, had been developed only in more recent times. Those were Darwin's decades, the idea of evolution seems to have pervaded even classical philology. A correspondent of *Nature* was eager to support the idea by his personal experiences, he himself having been colour-blind.

R. Schiller, *Between One Culture*, https://doi.org/10.1007/978-3-030-20538-6_1

The problem again surfaced after more than a century. In the years 1983 and 1984, *Nature* published a great number of ideas, arguments and counter-arguments on this subject. First in the series was a paper by two chemists. They realized that the Greeks drank wine usually diluted with water so they proposed that the spring waters of the area, often mildly alkaline, turned the wine blue. Thus, their idea was that the libation to Dionysus was a pH indicator.

Gladstone's idea regarding colour blindness was again proposed. Nowadays, this is easily refuted. On the one hand, phylogeny is much too slow to be compatible with the required change. On the other hand, let us visit the Acropolis Museum in Athens! One of the first items to be seen is the archaic statue of the sometime tyrant of the city on horseback, the mane of the galloping charger flies in blue. Going back in time, blue papyrus and blue sea with ships can be seen on the frescoes (Figs. 1.1 and 1.2), recovered from the ruins of Thera, which was demolished by a volcano catastrophe around 1600–1500 B.C.

Ideas based on psycholinguistics and anthropology were also considered. A linguist wrote that in different languages the number of names of basic colours

Fig. 1.1 Papyrus—Thera

Fig. 1.2 Sea port—Thera

varies between two and eleven. Present-day English can distinguish eleven colours, ancient Greek, however, could name only four. Also, translation could be difficult: the colours were not seen but called differently. The Pre-Socratic philosopher, Empedocles, wrote about colours; these writings, however, have been handed down to us in two contradictory fragments. One of them records that Empedocles defined only two colours, white being the colour of fire and red that of water. The other one states that since the philosopher distinguished four elements accordingly he knew four colours: white, black, red and yellow. But perhaps those interpreters are right who think that for the Greeks colours characterized rather the intensity of light than what present-day people call colour. To express this in the language of physics, we think of different frequencies, they thought of different amplitudes.

It might well be that a sharp-eyed philologist's approach is the best. He realized that Homer called the sea wine-dark always around sunset or early in the evening. During daytime, the Greeks sailed along the shore and moved to the high seas only when it was dark. It would have been dangerous to sail in the darkness among shoals and reefs. The sailor saw the sea as wine-dark when he made for the high seas. The philologist thinks the sea might have resembled present-day mavrodaphne wine.

One might think that such a cornucopia of ideas and discussions (and many times more, this being only a short summary) must have solved the question. That is not the case, Google presents a number of new references starting from October 2010.

As far as physics is concerned, the colour of the sea is more or less well understood. Tyndall was the first to recognize that colloid particles, which are much smaller than the wavelength of the light, strongly scatter light. The intensity of the scattered light depends on the wavelength, the short wavelength components of the sun's white light being scattered more effectively. Tyndall explained the green (!) colour of the sea by that phenomenon. Being an excellent experimentalist theory was not his forte, thus the mathematically correct theory of the phenomenon was developed by Rayleigh. That is the reason why what was originally called Tyndall effect now appears as Rayleigh scattering in the textbooks.

Several decades later the quantitative theory of the sea colour was approached on this basis. Four processes were considered: light absorption by water, light scattering by small air bubbles, reflection by fine silt and the reflection of the sky. The numerical evaluation of the corresponding equations necessarily rests on a number of assumptions since the processes depend on the momentary and local state of the water. The theory has not been widely applied. An empirical scale is used instead: the Forel-Ule scale consists of 21 standard colours to tell the changes of sea colour in the course of years and decades.

The space age has changed the type and amount of information for oceanographers. The sea around Crete indeed looks blue from space—well, mostly. Colour is greatly influenced by the winds (Fig. 1.3).

Fig. 1.3 Crete and the sea around it as seen by a satellite, July 28, 2003, natural colours. http://kkelly.apl.washington.edu/projects/aegean/

Oceanographers and climatologists seem not too much interested in the question of colour, satellites measuring mainly temperature. A geostationary satellite charged with colour measurements is being still planned and hoped for.

Perhaps trying to understand a question of literature along scientific ideas is in vain. Artists must be consulted. *Claude Lorrain*, the great French painter of the seventeenth century, was happy to populate his landscapes, be they real or imaginary, with biblical or mythological persons. Here is one of his pictures (Fig. 1.4). An episode of the *Iliad* is depicted in a most Baroque port: Odysseus returns captive Chryseis to her father. Let us look at the sea! The best word for its colour is—wine-dark. Obviously, Claude Lorrain never met Homer in person. But he met seas at noon and around sunset, in stormy weather and at calm and knew very well that the bluish light of the surface does not suit all events.

Another artist from later times might testify in a similar way. *Turner* lived around the turn of the eighteenth and nineteenth century in England. Several of his seascapes represent some episode of mythology. Here is for example the celebrated scene of the *Odyssey* with the hero mocking the giant Cyclops,

Fig. 1.4 Claude Lorrain: Odysseus returns Chryseis to her father

Fig. 1.5 William Turner: Odysseus mocks Polyphemus

Polyphemus, who hurdles huge pieces of rock towards the escaping ship (Fig. 1.5). The water waving around the ship trying to find her way among threatening cliffs and thundering stones is not friendly blue. This fateful sea is wine-dark.

Bibliography

Deutscher G. Through the language glass. London: Heinemann; 2010.

M'Master W. The causes of the colours of the sea. Nature. 1871;4:203–4.

MacNamara M. Homer's wine. Nature. 1984;307:590.

McClure E. Colour-blindness. Nature. 1878;19:5.

Pole W. Colour blindness in relation to the homeric expressions for colours. Nature. 1878;18:700–4.

Rutherfurd-Dyer R. Homer's wine-dark sea. Cambridge: Cambridge University Press; 1983.

2

Nessus' Blood

The mighty hero of Greek mythology, *Heracles*, had a painful and merciless death meted out by his own wife, *Deianeira*. The story, as it is related by *Sophocles* in his tragedy *The Trachiniae* (The Women of Trachis), is an appalling fabric of jealousy and revenge. And, I guess, also of some inorganic chemistry. At the beginning of the events, Heracles wants to cross a river with his wife finding there the mischievous centaur, *Nessus*, acting as the ferryman. The centaur has a simple way: he gathers up the passengers in his arms and takes them over. Doing the same with the wife he "*touched [her] with lusting hands*" midstream, something which is taken ill by the husband of wild rage. Taking his bow and arrow he inflicted a deadly wound to Nessus (Figs. 2.1 and 2.2). The dying centaur was deceitful with his advice given to the wife:

> *"Daughter of old Oeneus, listen*
> *to me, and you will profit from this voyage,*
> *for I will never carry any other.*
> *Take in your hands the clotted blood around*
> *my wound, in which the monstrous beast of Lerna,*
> *Hydra, once dipped his arrows of black gall;*
> *and this will be a love-charm for the heart*
> *of Heracles, so that he will not ever*
> *love anyone he looks on more than you."*

The tragedy opens with Deianeira receiving the news that his husband is going to take a new mistress to their house. The hero had never been flawless in his marital fidelity, but enough is enough—now it is time to make use of Nessus' blood. She gets hold of the well-kept blood, soaks a handful of wool

© Springer Nature Switzerland AG 2019
R. Schiller, *Between One Culture*, https://doi.org/10.1007/978-3-030-20538-6_2

Fig. 2.1 Heracles kills Nessos. Greek cup, Museum of Fine Arts, Boston

and smears over the husband's festive robe—she is going to send it to him because, if once he puts it on, he would not cast a look at any other woman anymore. (So it happened; the robe burnt away his flesh and in his torments he longed for death on a pyre. This, however, is not our business now.) Deianeira, shocked and frightened, relates to the women of Trachis standing nearby what she saw when smearing the robe over with the blood.

> *By chance I had thrown the piece of wool with which*
> *I smeared the robe into the blazing heat*
> *where sunlight fell; and as it warmed, it melted*
> *away to nothing, crumbling into earth*
> *exactly like the little particles*
> *of sawdust which we see when trees are leveled.*
> *It lies there still. And from the place it fell*
> *a curdled clot of bubbling foam seethed up,*
> *like the rich juice squeezed from the purple fruit*
> *of Bacchus' vine, when poured upon the ground.*

This description has a strange ring among the ancient lines. It sounds like a graphic account on an observed process rather than dramatic diction. The reader may have the impression that here Sophocles does not re-tell the myth,

Fig. 2.2 Heracles kills Nessos. Fresco in Pompeii, Casa del Centauro

instead he relates a personal observation—in terms of poetic expressions but with rare exactitude. As if he spoke about a chemical reaction seen with his own eyes. May this be the case? Could one propose a substance which attacks wool as described?

If Nessus's blood is something that can be identified as a chemical, there are three requirements this substance must fulfil.

1. It must have some connection with one or another element of the myth.
2. Only such materials and methods can be considered which were known, or at least could have been known in Antiquity.

3. If the Sun heats up this substance, it must attack wool so that bubbles form and only dust remains.

As far as the first requirement is concerned, Centaurs are known to have taken their origin from a region of Thessaly called Magnesia. Here, the mining of the mineral black manganese (pyrolusite by its scientific name, MnO_2 by chemical composition) had been going on since ancient times because it was an important additive in glass production as a decolourating agent. If black manganese is heated with saltpetre or potash in open air, the chemical compound potassium permanganate forms easily. Saltpetre and potash were known in Greece.

My colleague, János Balog after some reading in inorganic chemistry and after a number of experiments, finally mixed potassium permanganate with concentrated sulphuric acid obtaining a thick, oily liquid of the colour of clotted blood. Immersing a batch of wool into the liquid it did not attack the material when cold. However, at a temperature of about 30 to 40 degrees centigrade, it dissolved the wool under intense bubbling resulting into nothing else but some greyish dust. Attica's sunshine can easily produce the heat needed.

Obviously, the reaction depends on the composition of the liquid; very dilute or very concentrated solutions do not attack the wool, and there are compositions where the wool goes up in flames. There is, however, a broad range where one sees what was described by Sophocles.

There remains the question whether sulphuric acid was known in Greece and here the historians of chemistry do not agree. There is no known description, still some of the scholars think by indirect evidence that the Greeks could have or even must have obtained this substance.

Now we dare to surmise, and not without some evidence, that Nessus' clotted blood was nothing else but a mixture of potassium permanganate and concentrated sulphuric acid.

The ancient text adds something more, which is also telling. The blood was mixed with the gall of the hydra of Lerna. This mixture was kept by Deianeira. Can one identify gall chemically? Did the centaur's gall bladder contain sulfuric acid?

We thank Professor Zsigmond Ritoók for support and advice.

Bibliography

Balog J, Schiller R. Nessus' blood: chemistry in mythology. Acta Antiqua Hung. 2008;48:339–43.

Sophocles. *Trachiniae*. Robert Torrance, editor. Sir Richard Jebb, Translator. http://www.perseus.tufts.edu/hopper/text?doc=Soph.%20Trach.

3

Shelley Scientist

Once upon a time I read that there are three things we can gaze at indefinitely: the flames of fire, the waves on the water and the clouds in the sky. Poets make good use of this. Let us take the clouds—from Indian epic to *Wordsworth*, from *Homer* through *Pushkin* to *Swinburne* poems are densely packed with clouds. Description, image, simile, metaphor, allegory, symbol: one could compose a full meteorological style guide. The age of romanticism particularly delighted in clouds, in their tempestuous rage or their amiable golden edges. Seldom do they surprise the reader.

But sometimes they do. The great Indian-British-American astrophysicst, Nobel-prize winner *Subrahmanyan Chandrasekhar*, proved to be more discerning than most of us when it comes to reading poetry. In an important essay, he quoted the last stanza of *Shelley*'s poem "The Cloud".

> *I am the daughter of Earth and Water,*
> *And the nursling of the Sky;*
> *I pass through the pores of the ocean and shores;*
> *I change, but I cannot die.*
> *For after the rain when with never a stain*
> *The pavilion of Heaven is bare,*
> *And the winds and sunbeams with their convex gleams*
> *Build up the blue dome of air,*
> *I silently laugh at my own cenotaph,*
> *And out of the caverns of rain,*
> *Like a child from the womb, like a ghost from the tomb,*
> *I arise and unbuild it again.*

© Springer Nature Switzerland AG 2019
R. Schiller, *Between One Culture*, https://doi.org/10.1007/978-3-030-20538-6_3

Mary Shelley, who published this poem, thought she knew how these lines were conceived: "*They were written as his mind prompted: […] marking the cloud as it sped across the heavens, while he floated in his boat on the Thames.*" No wonder that, according to Mrs. Shelley, "*this poem bear[s] a purer poetical stamp than any other of his production*".

Chandrasekhar is of different opinion. Not that he did not consider this text to be the peak of English poetry. However, he knows, also following the ideas of King-Hele the physicist-cum-man of letters, that Shelley is a poet-scientist. The passage quoted above is a proof of that. In fact, all that is described here was unseen by the poet in his floating boat, it was not to be perceived by any means, its sound was not to be heard, or its scent to be smelled. There was only one way: it must have been learnt and understood. Reading the poem, we witness a particular wonder. With utmost poetical power, by making use of a very intricate and still highly harmonic form, creating a majestic scene, a genius of the English language describes—the natural circulation of water. Something what has never been seen and cannot be seen. Something that is a matter of diligent learning and clear understanding.

As written in a history of English literature, the poet "*at Eton […] hung entranced over the forbidden marvels of chemistry*". (What? Chemistry at Eton forbidden? In the age of Priestley and Dalton?) Chandrasekhar quotes *Whitehead*: "*Shelley's attitude to Science was the opposite pole to that of Wordsworth*" who considered science to be the murderer of Nature's beauty. Shelley "*loved it, and is never tired of expressing in poetry the thoughts which it suggests. It symbolizes to him joy, and peace, and illumination.*"

The history of literature draws a somewhat different picture: beauty is an "*intellectual*" experience for Shelley this being the spirit which acts in the Universe. "*And, in The Cloud, Shelley quits the guidance of Greek divinities and, with superb and joyous ease, makes myth for himself.*" Myth?

Certainly not science, since the poet made it clear that poetry, as opposed to logic, is not exposed to the active power of intelligence. Neither intention nor conscious resolve can help the poet. He is unable to recall knowledge wilfully. Knowledge of science even less. A poet of Shelley's stature can never compose a didactic poem. However, knowledge and understanding are nested in the poet's mind. In Shelley's case, that holds for the understanding of science as well.

A myth encompasses a total view of the world (as it was taught by the Hungarian scholar, Carl Kerényi). If so, a myth has been created in this poem. The myth of the scientific aspect of the world. The poem is unique also from this point of view, perhaps in the strictest meaning of the word. I wonder if there has ever existed or even there exists today, in an age of scientific world view, any poet, who was or is able to transform the (impeccable!) knowledge and understanding of Nature into poetry and music.

Perhaps it is not the poets' strength that is wanted. Their understanding and conviction of scientific truths, deeply rooted in their spirits, is called for.

Bibliography

Chandrasekhar S. Truth and beauty. Chicago, IL: The University of Chicago Press; 1987.

Hereford CH. Shelley. In: Ward AW, Waller AR, editors. The Cambridge history of English literature, vol. XII. Cambridge: Cambridge University Press; 1932.

Kerényi C. The gods of the Greeks. New York: Thames and Hudson; 1951.

4

Spectres Were Haunting Europe

"*I have nothing to retract. I adhere to my already published statements.*" This self-confident sentence was written by a great scientist not much before his death.

Presumably, the name of *William Crookes* (1832–1919) is less well-known than his most important invention, which is still in daily use. He was the first to build a cathode ray tube, which became the predecessor of all TV screens and monitors working on the principles of electron optics. This is, however, only one of the achievements of this great observer and experimentalist. He discovered the element thallium by making use of a then new method, spectral analysis, developed by Bunsen and Kirchhoff. Crookes immediately realized the great potential of the method, investigating the optical spectrum of a sample which contained the still unknown element. He constructed a gadget called radiometer, by now rather a toy or desk decoration. This is a paddle wheel in a low-pressure gas, one side of the paddles being shiny the other ones matte: if illuminated, it rotates around its vertical axis. Crookes thought this was due to the pressure of radiation—here he was wrong as it happened not infrequently with his explanations. As Hinshelwood wrote: "*it must be admitted that Crookes's analytical power hardly equalled his gift as an investigator of new facts.*"

He remained a keen experimenter, even in his later years. He was around seventy when radioactivity was discovered. Soon he realized that impinging α-particles, in those early days called p-particles, caused flashes on a zinc sulphide screen. This observation has become the basis of one of the most important methods of measuring radioactivity ever since. Later he conjectured that α-radiation consists of charged particles although the week magnets of his time were unable to deflect the rays.

© Springer Nature Switzerland AG 2019
R. Schiller, *Between One Culture*, https://doi.org/10.1007/978-3-030-20538-6_4

During the greatest part of his life, he lived on his own wealth working in his private laboratory. Now this genre of private scientist has been extinct (perhaps Lovelock was its last representative, who, among a good many other achievements, created the Gaia theory of life on Earth). England, however, acknowledged this type of activity around the end of the nineteenth century. Crookes became an influential personality, a grand protagonist of scientific life. Having been knighted, he was chairman of the Chemical Society for a longer period of time, was president of the Royal Society for two years and also of the British Association, which has played an important role in the popularization of science since the age of Davy and Faraday.

The Association was the venue of his important lecture about *"Radiant matter"* in 1879. Reading that title one would not expect that this clearly composed text based on brilliant experiments is just about gas discharge tubes. The figures and observations could find their places even in a present day textbook, some of the instruments being still in use in secondary school laboratories. Crookes analysed in great detail the effect of low gas pressures on all the phenomena which can be observed using elementary methods or even by naked eye. As a result of these clever experiments, he reached the conclusion that low-density gases of whatever composition undergo a phase transition into *"the fourth state"*, which is the *"radiant matter"*.

This was an erroneous statement but Crookes must not be blamed for that. His observations could be interpreted only in terms of the results of *Röntgen*, *Philip Lenard* and *J.J. Thomson* (who discovered the electron), which were at hand only some 10 years later. It is the final conclusion of the lecture which makes the present day reader bemused. *"Some properties of the radiant matter is of material nature, similar to this table, whereas its other properties almost assumes those of the radiant energy."* This was perhaps tolerable in those years. Also Ostwald dealt with the duality of matter and energy because the notion of matter was limited by them to entities which tilt the pair of scales. The continuation, however, is difficult to be tamed. *"Here we have reached the borderline where—it seems—matter and force get mutually transformed, the shadow world between the Known and Unknown […] I think here are the final realities to be found."*

Shadow world and final realities … it would be reassuring to give them a brave meaning but this is a vain effort. It is known that Crookes had been committed to Spiritism already five years earlier. He was in a large and good company. The movement, belief or fallacy, let it be called by any of these names, had been blooming for decades. As far as we know, it was born in Hydesville, New York, U.S. in 1847. Two young girls, Margaret and Kate Fox, heard curious noises, clatter in the house. Kate, twelve years of age, asked the invisible noisy Powers to clatter in the same way as she clapped with her fingers. The Powers proved to be obedient. The girls then started a conversation with the spirits, the answers having been understood by the number of clatters. The haunted house and the teens chatting with spirits soon rose to national fame.

Thirty years later, several thousands of believers of Spiritism lived in America, in England and in Continental Europe. National organizations of Spiritism were established, their first journal was published in England in 1855, their national congress in America decided to discard the rules of all Christian religions and to close the Sunday schools in 1866. The movement was successful in the highest ranks. In Lincoln's time, séances were held in the White House, Queen Victoria tried to contact his late husband, Prince Albert. Prime Minister Gladstone was no less a spiritualist than Conan Doyle who created the figure of the champion of sober logic, Sherlock Holmes.

The second half of the nineteenth century was one of the golden ages of science, a classical period of exact research. And also the age of apparition? Something similar happened hundred years before when the contemporaries of Lavoisier and Priestley tried to make gold. Strict science seems not to limit twilight and misconceptions; on the contrary, it enhances their prevalence. Science was seen to bring about things which had been unbelievable. Then why not to believe in making gold or conversation with the other world? The work of Faraday and Maxwell revealed a system of phenomena which had nothing to do with tangible objects the weights of which can be measured on a scale. The electromagnetic phenomena were beyond any "common-sense" expectation, as far as the limits of common sense rest on objects which are visible to naked eye and can be kept in hand. The propagation of the electromagnetic field takes no time (in the traditional sense of the word), a magnetic rod exerts its influence from a distance—these phenomena are beyond belief. Even the clattering of the spirits echo a technical idea. Morse patented the electric telegraph in 1838, which—after heavy protests—was set in operation a decade later. It gave information from non-tangible persons in invisible distances.

This feeling arose also in literature. In Dickens' short story, "The Signalman", a supernatural, ominous Appearance figures, intertwined with Morse messages and the irrational ringing of an electric bell.

Scientists took seriously this attack of undemanding irrationalism and stuck-in-mud transcendence. Crookes undertook the task to investigate the effects of Spiritism in strict scientific terms. This seemed to be an excellent choice in view of his multifarious activity in chemistry and physics and the unquestionable exactitude of his experiments. At the outset, he made clear that he was completely unbiased, being convinced that science could not yet have fathomed the depths of all physical phenomena.

However, his first report was greatly disappointing and was fiercely rejected by the community of scientists. After having attended a spiritualist séance, he wrote that from this time on common sense and sensory impressions contradicted each other in his mind. He experienced a number of effects, regarded as occult, although he tried to analyse the events objectively and without prejudice.

Later he surrendered completely, although still made use of the tools of experimental physics. He was acquainted with fifteen years old Miss Florence Cook who was hailed as wizard-medium all over London. The appearance (or apparition) was the following. As soon as Florence had fallen into a trance in a faintly lit recess, there appeared another girl stating that she was a spirit, her name was Katie King and was the daughter of Henry Morgan, the pirate. (One thing is sure: Morgan had been a real person living between 1635 and 1688.)

Crookes tried to elucidate two questions. First, he wanted to make sure whether there are two persons. Thoroughly he measured their heights, noticed a bruise on the neck of one of them, the pierced ear of the other, made photographs—no doubt, two distinct persons were present. Then, he was to prove that no flesh-and-blood person could enter the room but there was a spirit on the scene. For that purpose, he built a simple motion detector, which consisted of an electric circuit connecting the bracelets on each hand of the medium. Had she taken off the bracelets the circuit would have given a signal. This, however, never happened. On another occasion, he asked the spirit to immerse her hands in a jar full of an electrolyte solution. If it had not been the spirit but the medium, the motion detector would have given a signal but the attendees of the séance had no such experience. (These experiments proved, if anything, that two persons were present, but not that one of them was a spirit.)

The series of experiments took three full years after which Katie's spirit finally disappeared. Crookes became fully convinced by his own experiments that what he saw was real—whatever that means in this context. As far as formal reasoning was concerned, his argumentation was flawless. If his opponents state that his experiences are against the laws of nature, they regard the debate already finished. But this has always been the method of reasoning by which people have tried to refute any great discovery—he argued.

On the other side, a popular science journal clarified its position by saying that there exists a basic antagonism between Crookes' Spiritism and the journal's materialism—let them leave each other alone. John Tyndall, a celebrated scientist-polymath of the day made a reproach against Spiritism in view of lacking uniform reproducibility. "*When science appeals to uniform experience, the spiritualist will retort, >>How do you know that uniform experience will continue uniform? You tell me that the sun has risen for 6000 years; there is no proof that it will rise to-morrow<< [...] The drugged soul is beyond the reach of reason.*"

Credulousness, fallacy, expectation of miracles? Harry Houdini, the Budapest-born American conjurer enjoyed unmasking spiritualist trickeries in his later years publishing a book entitled *A Magician Among the Spirits. The conjurer tricks* he presented were completely convincing in this aspect. Thus, the Editor of *Scientific American*, who offered a prize to any medium who could prove the real existence of spiritualist phenomena, stated that no effect has ever been observed which cannot be produced by natural means. This happened as late as 1925.

Crookes did not live to know this verdict. The argument between science and superstition has not been concluded, anyway. In his last years, the scientist, having lost his beloved wife after sixty years of marriage, was more convinced in his belief than ever.

Nevertheless, this has been seriously refuted already in his lifetime. In their later years, the Fox sisters, founders of the movement, admitted that they had lied in their childhood, just to tell later that this admission had been a lie. None of their statements can be taken seriously both of them having become incurable alcoholics. But there is something else.

When Katie King's spirit made her (its?) regular appearances, the medium Miss Cook got married. After the ceremony, the young husband, Captain Corner, found the spirit and the medium in the same bed. This story was set down in Queen Victoria's bashful era, thus ... the rest is silence.

Bibliography

Beichler JE. Either/or: spiritualism and the roots of paranormal science. Either-Or Spiritualism.htm (n.d.)

Crookes W Strahlende Materie oder der vierte Aggregatzustand (übersetzt.Gretschel H). Leipzig: Quandt und Händel; 1907.

Der Okkultismus der Avantgarde um 1900.htm

Hinshelwood CN. W. Crookes. In: Dictionary of national biography; 1927.

Pement E.: New Age Part II: channelling, messages by remote control. cornerstonemag.com (n.d.)

Part II

Alchemy

5

A Puzzling Tale of Gold

Gold cannot be made. The Priest, a simple good soul, must have learnt this to his own cost. He is one of the characters in *The Canon's Yeoman's Tale* of the *Canterbury Tales*, a chain of stories written mostly in verse by Chaucer around the second part of the fourteenth century. The story line is a pilgrimage to Canterbury, the company of pilgrims telling each other stories, entertaining ones and salutary ones, just to make the long ride more lively and amusing. This pilgrimage must have been somewhat similar to a present-day package tour.

The Canon, the protagonist of the tale we are now referring to, was not a member of the company; the sad history of his fraud was related by one of the pilgrims to the rest of the company doing it in such a lively style and with so many details that some experts think the poet might have had some similar experience. The story is a simple confidence trick. The Priest, gullible as he was, was approached by the Canon who presented himself as an expert alchemist being able to convert base metals, like copper or mercury, into silver. He made the priest lay a big fire, put a crucible in the middle full at one time with copper, the other with mercury, then

> [t]his false canon – the foul fiend him fetch! –
> Out of his bosom took a beech-wood coal,
> In which all subtly he had bored a hole,
> And put therein silver filings from the scale,
> An ounce, and sealed it was, without fail,
> That hole with wax, to keep the silver in.
> ………………………………………..
> And when the canon's beechen coal
> Was burnt, all the metal from the hole

© Springer Nature Switzerland AG 2019
R. Schiller, *Between One Culture*, https://doi.org/10.1007/978-3-030-20538-6_5

Into the crucible flowed down anon
– For so it had to do, as stands to reason,
Since laid so levelly above it was.
But the priest knew naught of it, alas!
He thought all the coals equally good,
For of the trick he nothing understood.

After all that a pan full of water was taken, the concoction was poured into the pan, making the simple minded Priest find some silver powder together with a sheet of silver since that too was smuggled into the pan by the Canon. Thereafter, the mysterious prescription was sold to the Priest. "*You must pay forty pounds, so God me save!*" That was quite a neat sum; Chaucer, already a well-known poet and a clerk at the Court, was paid twenty pounds per year by King Richard II.

The lesson is clear: "*A man easily learn, if he owns aught, /To 'multiply', and bring his wealth to naught.*" The lesson is far from clear. The tale is told by the Canon's Yeoman. Both he and the canon, looking miserably poor, join the company midway. First, the Yeoman boasts his master's power in alchemy, then he explains that the Canon is poor only because he is much too clever. They dwell in some dark and remote nook, a place for outlaws. Are afraid of the law or are they just very miserable? Does the Canon deceive the gullible ones as the Yeoman says? "*He has betrayed folks many a time; Of his falseness it troubles me to rhyme.*" Or is it grave self-deception?

But the science runs so far before
We cannot, despite the oath we swore,
Overtake it; it glides away so fast.
It will leave us beggars at the last.'

The Canon, frightened and hurt, slapping his horse sped off. It is only then that the Yeoman starts with his tale. However, the tale is not about his own master, although there are honest canons, too. (Whoever maintained the contrary?) The Canon cannot be a simple cheat either, even his servant being a trained scientist. After having quoted the *Rosarium* by *Arnoldus de Villanova*, a basic work in alchemy, he talks about *Hermes Trismegistos*. As if the text suggested that master and servant, experimentalists nagging at each other while busying with crucibles, bellows, dephlegmators and retorts, one of them pointing to the wood the other to the clay pot as a cause of the failure, are no swindlers rather ill-trained, poor experts. They ought to know that mercury and sulphur can exert their effect only in concert! The mysterious sentence must refer to something like that: "[*Hermes*] *says that the dragon, if you please,* / *Does not die unless he in turn is slain.*"

Nowadays, this pair of dragons is denoted in a prosaic way as Hg and S. Symbols and language have changed during the centuries, and this makes it even more difficult to understand the alchemist texts. But here is something more. Chaucer quotes "*Plato*", obviously not the Athenian philosopher but some unknown, legendary alchemist. Answering the disciple's questions, he calls the Philosophers' Stone on one occasion *Titanos,* on another *Magnesia,* then again a "*liquid that is made* […]/*Out of four elements*". Finally, he rejects the question:

> *'Nay,' quoth Plato, 'It is a secret, still!*
> *The philosophers are sworn every one*
> *To reveal the essence of this to none,*
> *Nor write it in a book in any manner,*
> *For to God it is so precious and dear*
> *That he wishes not its discovery,*
> *Save where it is pleasing to his deity*
> *To enlighten men, and thus to defend*
> *The truth from others; lo, this is the end!'*

That is also the end of what Chaucer tells us. It seems to be certain that if the reader is erring in the story that is not because this excellent narrator cannot draw a clear picture. Did the trickster perform one experiment or two? Why did he use both powder and sheet of silver? Who was the Canon who fleeced the Priest? Are all alchemists crooks? According to its content, the tale is about a vulgar fraud. According to its mode of presentation, it is about some mystery which is hidden for common mortals.

Gold must not be made.

Bibliography

https://en.wikisource.org/wiki/The_Canterbury_Tales/The_Canon%27s_Yeoman %27s_Prologue_and_Tale

6

Alchemy: Its End

Mysticism and fraud—these are the words that come to one's mind when talking about alchemy. Whatever the spiritual basis and the practical aim of the alchemists, all their activities were closely connected to laboratory practices, that is, their mysticism had an experimental footing. They wanted to effect a change in nature, particularly in the nature of metals, called transmutation, a process which repeatedly defied well-known metallurgical methods; hence, they were compelled to turn to the black arts. They knew only too well that without heavenly support, their endeavour would certainly fail; thus, they were advised to fear God and conduct a virtuous life. This laboratory wizardry makes practical processes and magic symbols appear intertwined, and it is often difficult to tell whether an expression is to be understood in its direct meaning or it hides and perhaps discloses some secret. Most often all the three are true at the same time.

Vocabulary and graphical representation makes understanding even more difficult. Substances are referred to by graphical signs, processes by pictures showing persons or events. Whereas self-inflicted secrecy prevented the authors to give clear explanations, understanding is made even more difficult by their practice, probably rooted in their mystical way of thinking, that one and the same substance or treatment is often called by different names, referred to by different signs. The more important an item was the more names and signs were assigned to it. At a later time, when rational thinking was about to take hold Sommerhoff's *Lexicon Pharmaceutico-chymicum*, published in 1701, gave a list of rational names and their alchemist signs. Salpetre solution, which obviously played little role in the prescriptions, had only one sign. Antimony

R. Schiller, *Between One Culture*, https://doi.org/10.1007/978-3-030-20538-6_6

and arsenic, however, were important substances in the magisterium, thus they figure with about forty symbols, the same being true also for gold.

It is not only about names. Literal meaning, esoteric reference and the magical effect of tools and treatments often appear inseparably. Take for example the pelican, a laboratory vessel, most often used in the alchemist's kitchen. Similarly to the role of our present-day reflux coolers, the pelican prevented a liquid from evaporation when heated even over an extended period of time, the process itself being called *circulatio*. During heating, the base metal (e.g. copper) was intended to be destroyed so that after its death, it can be reborn in the form and essence of a noble metal, silver or gold. So death and resurrection are connected to the vessel, which borrowed its name from a bird with a large bill and a long, bent neck. The shape of the vessel resembles the bird. According to an old belief, stemming from Alexandria and enriched with Christian elements in Byzantium, this bird although loving its nestlings kills them, but after three days it rips up its own body and with the drops of its blood it effects their resurrection. Thus, the bird became a symbol of Christ who resurrected and promises resurrection to all.

Did the vessel obtain its name only by its shape or also by its mystic role? What is the metaphor of what: the vessel of the bird or the bird of the vessel? There is little relief in rational thinking (Fig. 6.1).

Paracelsus, the first real iatrochemist, who was active in the first half of the sixteenth century, decided to synthetize medicines instead of just finding

Fig. 6.1 Pelicans. Left hand side: the alchemical vessel, right hand side: relief from the Münster Cathedral

them. The first to realize that medicine and poison are not necessarily different in composition but rather in dose. Paracelsus was a diligent experimentalist and at the same time a mystic thinker who firmly believed in astrology and had an approach to alchemy very different from that of his predecessors and contemporaries.

No doubt, he believed in alchemical methods and aims, but this belief was closely connected to his own ideas of elements in nature, which, in its turn, was based, at least partly, on his own experiments. Destructive distillation was an important method of those times. The substance to be investigated was decomposed by heating, and the products of decomposition were cooled down and collected. That is similar to the process taking place in a charcoal kiln.

The belief, now considered utterly naïve, was that the substance under study contained the products with the same composition and in the same aggregate state as they appeared in the collector under the cooler. The aggregate state was thought to be conserved during chemical changes as nowadays mass or electric charge are known to be conserved. This belief must have suggested to Paracelsus, not without predecessors, the idea of *Tria Prima*, the Three Primes. Three substances were thought to be the elements or rather the constructing principles of every material: *Sulphur, Mercury* and *Sal* (salt), sulphur representing the ability to burn and evaporate, salt the inertness of the unreactive solid and mercury the fluidity with the unexpected appearance and disappearance of a nimble liquid. Most probably, when talking about his primes, Paracelsus had not the real, tangible materials in mind, rather ideas of their behaviour.

It is difficult to believe that Newton, *humani generis decus*, the ornament of the human race, father of classical mechanics, whose strict axiomatic reasoning has been the paradigm for science and scientists ever since, had anything to do with alchemy. But he had.

According to his contemporaries, he was a thoroughly trained, conscientious experimentalist working hard in his kitchen. His writings in this field, consisting of notes, extracts of texts and laboratory diaries, probably surpass his scientific works in volume. Nevertheless, he published none of them. Even the extent of these manuscripts suggests that he spent more time with alchemy than with physics and mathematics together. How to understand this? Did he think indeed that the alloying of antimony with gold or silver is a mystery as it is described in an extant manuscript entitled *Clavis*, The Key, a work attributed to Newton? Was alchemy the key that unlocked Newton's heart?

In an unpublished paragraph of the Preface of the *Principia*, he made clear that his desire was to extend his mechanical studies to the structure of materials. "[…] *the motions of smaller bodies depend upon certain smaller forces just as the motions of larger bodies are ruled by the greater force of gravity. It remains therefore that we inquire by means of fitting experiments whether there are forces of this kind in nature, then what are their properties, quantities and effects."* This is a complete research programme for microphysics.

A paragraph of his *Opticks* is almost prophetic. *"Have not the small particles of Bodies certain Powers, Virtues or Forces by which they act at a distance, not only upon the Rays of Light for refracting and reflecting them, but also upon one another for producing a great part of the Phaenomena of Nature?* So, his conjecture was that optical phenomena are closely related to forces acting between particles which the substances consisted of. Some two hundred years after his death, quantum mechanics completed this programme. The primary aim of the theoreticians, Heisenberg and Schrödinger, was to understand the optical spectrum of hydrogen. The treatment of this problem enabled them to reveal the basic laws of microphysics, which were soon applied to the interactions of atoms. That was the way the physical nature of the chemical bond was discovered. Exactly as Newton predicted, starting from the phenomena of light ending up with the forces between the "small particles of the Bodies".

This being beyond the realm of Newton's mechanics, he necessarily could not achieve his own aim. Was this the reason he did not publish any of his alchemical papers?

Alchemy was not reserved for geniuses; it was pursued by lower grade researchers as well. Obviously missing Newton's insight, their down-to-earth results also reflect the spirit of the age when critical reasoning contrasted with mystical approaches. The idea of finding "mercurial water", a kind of water which was hoped to be obtained from mercury and would act as Tincture in the Great Magisterium of making gold, had always haunted the adepts' minds and was still alive in the seventeenth century. George Wilson, a serious worker of this generation, tried to obtain that substance by heating and distilling mercury. Indeed, he found a small amount of liquid but soon he realized that the water did not come from mercury but from the still which was not thoroughly dried. According to the spiritual atmosphere of his age, he even put down his observation; Proto-alchemist Hermes Trismegistus would never have done anything like that.

Making gold is obviously impossible for the present-day chemist, whereas it was not for the alchemist. But if possible, then why hopelessly difficult? Psychology has tried to find the answer. Carl Gustav Jung, a most influential psychologist of the first half of the previous century, started his professional career as a devoted follower of Freud. However, they parted soon because Jung, prone to esoteric thinking, was unable to tolerate Freud's rigid, nineteenth century positivism. In view of his readiness to digest mystic ideas, Jung studied alchemical literature with great diligence trying to find the alchemists' psychological motives. His main tenet was that the real meaning of the alchemist's work was a constant fight with his own unconscious, all the cryptic expressions, symbolic descriptions, the annoying inconsistency in wording were signals given by the unconscious which was reluctant to reveal its own nature.

Alchemist and his unconscious fought in the laboratory where the researcher could project his identity onto the process in the alembic. As if he had seen his own face reflected by the mercury in the bottom of the vessel. Chemistry is a perfect area for such an encounter. Although a chemical experiment is a process of nature, open to unbiased observation, still any reaction might often go astray because of uncontrollable external conditions. Failures are frequent, and their explanations are various. Therefore, this is an excellent battle ground for the unconscious which wants to remain hidden and at the same time to become exposed.

Given the important role that chance plays in this fight, it is little wonder that among the Primes of Paracelsus the main actor in alchemy was Mercury, the unforeseeable, untamed, omnipresent and always disappearing idea of a substance. It is often represented as Melusina, the mermaid who left her watery home for a husband, only to leave him when he revealed the secret of her body (Fig. 6.2). Jung calls her an *"elusive, deceitful and malicious elf that drives the alchemists into desperation"*.

Fig 6.2 Melusina representations in an alchemy text

The age of alchemy over, Melusina appeared in the literature of the twentieth century to represent the malice of chance. Josef K., he protagonist of *The Trial*, a novel by Kafka, was accused, tried and finally executed on grounds of unknown charges in the course of unforeseeable events. Perhaps the only person who gave him a sympathetic piece of advice was the servant girl, Leni, by saying *"you have to confess"*. Her hand was of a curious shape: *"She spread the middle and ring fingers of her right hand apart from each other. Between those fingers the flap of skin connecting them reached up almost as far as the top joint of the little finger."* It was like a web of a water-fowl. Or of a mermaid. And when K. left, he was soaking wet—Kafka says there was a heavy rain in Prague on that day.

It might be mere chance that Kafka lived for some time in the Prague Golden Lane (*Zlatá ulička*) where, according to legend, the alchemists to Emperor Rudolph pursued their art.

Bibliography

Beretta M. The enlightenment of matter: the definition of chemistry from Agricola to Lavoisier. Canton, MA: Science History Publications; 1993.

Dobbs BJT. The foundations of Newton's Alchemy. Cambridge: Cambridge University Press; 1975.

Jung CG. Paracelsica. Zürich: Rascher; 1942.

Jung CG. Psychologie und Alchemie, Gesammelte Werke XII. Olten: Walter; 1972.

Jung CG, Pauli W. Naturerklärung und Psyche. Zürich: Rascher; 1952.

7

Alchemy: Apologies

The age of reason, of pure ideas in science, superstition disappearing—that was the eighteenth century. At least that was the way how its best minds regarded their own age and also how we are ready to think about them. All these seem to be true, nevertheless, something else is true too. That century saw the last flourishing of alchemy. This is difficult to understand since alchemical ideas and practices are in stark contrast with everything that the century nurtured, nourished and boasted of. Alchemists knew this very well, and while sticking unwaveringly to their pre-scientific belief, they tried to find arguments for the scientifically minded public in order to recruit new supporters or at least to dispel the misgivings of the public.

Sándor Bárótzi, a member of the Hungarian Noble Guard to Empress and Queen Maria Theresia in Vienna, became a believer and practitioner of alchemy in his later years. More than that, he also played an important role in Hungarian literature through his translations. His two activities appeared sometimes in concert by translating an anonymous French novel, entitled *The Adept of Our Days or The Real Secret of the Free Masons*. Clearly stating that the novel is mere fiction he nevertheless composed an introduction in order to publish and defend his own conviction. He must have understood the general attitude of most of his readers in the time of the Enlightenment towards mystical and irrational ideas, so he felt compelled to make use of scientific achievements and technical results in order to convince the public.

Do you believe the acus magneticus (magnetic needle) *to turn always toward the Pole, or the magnet to pull iron and steel with such a power that one can tear it away only with force? Or a piece of ordinary steel having been rubbed with a magnetized steel to get an attracting power as large as that of the other one? [...] Do you believe*

© Springer Nature Switzerland AG 2019
R. Schiller, *Between One Culture*, https://doi.org/10.1007/978-3-030-20538-6_7

that if it were possible to form a row of people, hand in hand, from Vienna to Constantinople and the first one in Vienna would be electrified, the last one in Constantinople would feel the shock in the same hour or perhaps in the very moment? [...] Would you believe the lightning's path to be lined out in order to make it hit just at a given place and not at any other one? I bet you would not! Still Franklin found a method to impose a limit upon it. [...] Would you have believed even just fifteen years ago that one can make a voyage in air? [...] Mongolfier and Blanchard fly up and take their route in air from England to France.

I must admit I cheated somewhat with the above text by cancelling statements which, contrary to the author's time, nowadays are regarded neither scientific nor results (like those about animal magnetism). But those I have copied are also held true by us. The gist of the argument is clear: while these results contradict common sense and therefore are difficult to believe, they are still known to be true. If so, why not believe the alchemist's statements when the only argument against them is their improbability.

This way of argumentation is characteristic of an age in which science enjoys great authority. The author, or his source, presents us with the method of pseudo-science in a scientific age. One could simply discard science and refer to magic, trance or revelation—but no! This in Newton's or Lavoisier's age is not viable any more. And not only because no one would believe the statement, including the person who makes it. The alchemist of the eighteenth century, and all his pseudo-scientist successors until today, yearns also for intellectual rank and standing. And this can be obtained only from real, well-established science. Its methods are unknown to the pseudo-scientist, its actors are despised by them with envy and its results, however, are astonishing enough to make the bed for other (false) astonishments.

Thus, there is no point bearing a grudge against poor old Colonel Bárótzi. He was one of the firsts among those who wanted thorn apple to be improved in the garden of science. The weed stays what it is, but, one hopes, the treatment does not do much harm to the garden.

A good century later physical chemist Wilhelm Ostwald approached a similar question. *"A thing appears as mystic only if it is not well understood"* he claimed as, indeed, a scientist should. However, this is the reason why *"mystic feelings make a part of the primordial substance of our inner life"*. There is an enormous amount of material in the past of humanity and of our personal lives as well that we are unable to understand if only because of our ignorance. That is the reason why Ostwald feels some sympathy for the repeated resurrection of mysticism at the beginning of the previous century, although his scientifically trained mind is shocked by people who re-read not only Paracelsus but also the Shakespeare contemporary Robert Fludd, the Rosicrucian alchemist or the spirit-seer Swedenborg. The history of human thought has taught him that once the power of the generally accepted world view is exhausted and is unable to explain new phenomena, everyone will face questions which are impossible to be understood by the old principles. The easy answer to this challenge is the abandoning of rational thinking and an eruption of "*deep mystic feelings*". Hence, let's take courage! The development of science has always been accompanied by the ebb and flow of mysticism, since "*mysticism has always been only a transition to the new rational views*". That is the way how the human soul prepares to absorb the intellectually new.

Nowadays, Ostwald's thesis would be expressed in different words, by saying that a basic change in the scientific paradigm inevitably goes hand in hand with a revival of irrational thinking. I would not dare either criticise or comment upon this psychological and historical explanation of the contemporary re-birth of mysticism. The fair and open-minded assessment by the strict rationalist Ostwald of mysticism and mystics is certainly most inspiring. At present, the problem seems to be more pressing than ever. When rational thinking is menaced by a host of pseudo-, para- and other non-sciences, we had better have recourse to Ostwald's peaceful optimism.

Bibliography

Ostwald W. Forderung des Tages. Leipzig: Akademische Verlagsgesellschaft; 1911.

Part III

Poems and Science

8

Ignorant Doctor Faust

"I HAVE, alas! Philosophy, / Medicine, Jurisprudence too, / And to my cost Theology /
With ardent labour, studied through. / And here I stand, with all my lore, /
Poor fool, no wiser than before. / Magister, doctor styled, indeed, /
Already these ten years I lead, / Up, down, across, and to and fro, /
My pupils by the nose,--and learn,/ That we in truth can nothing know! /
That in my heart like fire doth burn."

These are the words of the scholar, Doctor Faust, disappointed with his own book-worm wisdom, uttered at the beginning of Goethe's tragedy. It has been for two hundred years that readers and theatre-goers have felt sympathy with the disillusioned wise man who is unable to find any consolation in all of his amassed knowledge. Indeed, it is easy to understand why he escapes his stale study with diabolic assistance.

As a matter of fact, Faust was a real person living around the turn of the fifteenth and sixteenth centuries. His fame of a satanic magician was due to a popular book of that period, which also served as a source for Goethe. The flesh and blood Faust was a contemporary of Paracelsus and many of the Faust commentaries are eager to list the motives and loci of the drama with references to or quotations from Paracelsus. But Faust, the real person, was a contemporary also of Agricola, the great geologist, metallurgist and miner. Goethe was well-versed in crystallography and mineralogy, he even conducted independent studies in these fields. Nevertheless, Agricola's sound, inquisitive mind does not appear in the play. Goethe himself being a contemporary of Lavoisier had an up-to-date knowledge of chemistry. Moreover, he made use of a chemical metaphor in his novel entitled *Elective Affinities (Die*

© Springer Nature Switzerland AG 2019
R. Schiller, *Between One Culture*, https://doi.org/10.1007/978-3-030-20538-6_8

Wahlverwandschaften) in order to bring home the underlying idea of the work. Despite all this: no Agricola, no Lavoisier throughout the play. On the contrary!

Soon after his first monologue and having finished a barren talk with his stupid famulus, he addresses his laboratory equipment idling on the shelves:

> *Ye instruments, forsooth, ye mock at me, / With wheel, and cog, and ring, and cylinder; /*
> *To nature's portals ye should be the key; /*
> *Cunning your wards, and yet the bolts ye fail to stir./ Inscrutable in broadest light, /*
> *To be unveil'd by force she doth refuse, / What she reveals not to thy mental sight, /*
> *Thou wilt not wrest me from her with levers and with screws.*

It is curious how the great scholar Faust considers scientific understanding of Nature. Disillusioned by all the tools and methods of the laboratory, he believes spontaneous revelations to be the only source of knowledge, since "levers and screws" will fail with certainty. Nature must be observed only, any coercion would be completely in vain. Simply said, there is no need of experimental science. Is that the scope of Doctor Faust's wisdom and knowledge? Why does not the mineralogist contemporary of Faust or the chemist contemporary of Goethe come to the author's mind? These people realized that experimental work is an evident and indispensable method for revealing the laws of nature.

It seems to be odd how Goethe considered science a century after the birth of classical physics and around the early development of chemistry. He studied nature with passion and insatiable interest, inanimate minerals and living organisms, plants or animals, as well. By visiting mines, he found new minerals; by making dissections, he discovered the human incisive bone (intermaxillary bone) and developed theories about a primal animal, the ancestor of every present-day animal, and a primal plant, the ancestor of every present-day plant (Fig. 8.1). However, most of his time spent on the riddles of nature was devoted to optical studies, to the sincere regret of the sensitive hearts among the readers of his poetry.

Farbenlehre (The Theory of Colours) is the title of the voluminous work, the aim of which was to present a description of the physical, chemical, biological, psychological, ethical and artistic aspects of light and colours. The author of this work greatly despises the methods of the physicist who performs experiments. As Faust disdains levers and screws so refuses Goethe, as an investigator of optics, Newton's prism. His stance is made clear at the beginning of the work.

> *Newton thinks that every colourless, white light, the Sunlight in particular, contain*
> *a number of lights of different colours whose superposition brings about white light.*

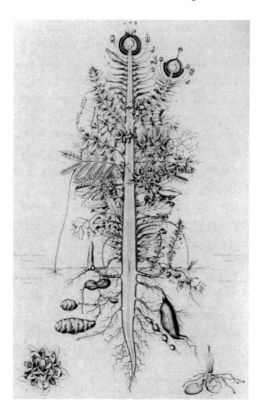

Fig. 8.1 The primal plant as imagined by Goethe (Woodcut by Pierre Jean Francois, 1837)

In order to reveal the coloured lights he imposes definite conditions to the white light using refracting materials which deflect the ray from its straight path; even that is made in no simple equipment. He gives a number of different shapes to the refracting materials with which the space is set up; he limits the light to small apertures and thin slits and having cornered the light by thousands of different methods he maintains that all these conditions had no other effect but to reveal the properties and abilities of the light by opening up its internal parts and making its content visible.

The procedure which we regard as the only method of cognition, an opinion which has been, no doubt, effected by Newton, is completely inviable for Goethe. Instead, he lists and systematizes innumerable sensory impressions and observations, trying to create a unified colour theory. Obviously, he was as successful as with another endeavour of his, where he tried to reveal the primordial and ancient form of all existing plants and found a thistle on the side of the road.

As if Paracelsus, the alchemist, had fought Newton, the mathematician. Most probably Goethe was not aware of Newton being a sedulous alchemist himself. No wonder, Newton concealed his alchemist activity, not making it known that he probably spent more time with and made longer notes about alchemy than physics and mathematics together. Nothing was published about alchemy during his lifetime. Would Faust, who held regular conversations with spirits, and Goethe who tried to understand the nature of colours by instinct alone, be placated towards Newton's way of thinking if they had known that here the laws of motion and mathematical analysis lived peacefully together with the alchemists' hazy wording?

Perhaps even that would not have reconciled the poet's and the scientist's world views. Heisenberg reminds us of a particular feature of the naturalist Goethe's way of thinking. He quotes a passage from the poet's letter: *"And it is exactly the greatest of all harms the new physics commits that it separates man and experiment. [...] And computations are treated in a similar way. There are numerous truths which cannot be computed and again there are a good many of them which cannot be decided by experiments."*

Most probably, Goethe knew only too well that nature can be understood only through abstraction; he himself, however, was not willing to go along that path. Seeing it to be much too dangerous, he refused to renounce the non-mathematical part of man's wealth. He found Faust's old-fashioned misinformation to be a small price to be paid for the salvation of this treasure.

Bibliography

Goethe: Faust, Part 1 http://www.gutenberg.org/cache/epub/3023/pg3023.txt
Goethe. Farbenlehre, Goethes Werke in zwölf Bänden, Zwölfter Band. Berlin: Aufbau-Verlag; 1968.
Heisenberg W. Physikalische Blätter. 1968;24(193):241.

9

Hölderlin, Blake: Newton

The picture and the poem were born at about the same time, in the last decade of the eighteenth century. The poem was written by a young man in Germany, solitary and most sensitive; the picture was created by an English painter and mystical poet, half a generation the young man's senior. *Hölderlin*, the German, a younger contemporary of *Goethe* and *Schiller* was probably the greatest and certainly the most ill-fated genius of German romanticism. The painter who created his poems and prophetic books, together with his graphic art, produced and printed with a peculiar technique, was called *William Blake*. Hölderlin, admiring the spirit of the Greeks and enchanted by a great and unrequited poetic love during the short creative period of his life, spent his last decades in the misery of schizophrenia. Blake, having toiled with great visions during his long life, expressed all his inner experiences in a symbolic language, which made him perhaps the greatest poet of the early Romantic movement in England. Certainly, the last thing one would expect of either of the two Romantic poets is to be inspired by the strict ideas of Science. Still, Hölderlin celebrates Kepler and Newton in an ode, and Blake paints Newton's figure or, to be more precise, paints a figure stating that it represents Newton. It is not easy to understand.

Perhaps Hölderlin's case is the simpler. He was born in Lauffen, a small town in Schwäben (Latinized as Suevia), a region towards which he nurtured patriotic feelings. This local pride led him to Kepler's person who was born in a nearby township called Weil. The ode glorifies both Kepler and their common native land. The last lines are as follows:

© Springer Nature Switzerland AG 2019
R. Schiller, *Between One Culture*, https://doi.org/10.1007/978-3-030-20538-6_9

Mutter der Redlichen! Suevia!
Du stille! dir jauchzen Aeonen zu,
Du erzogst Männer des Lichts ohne Zahl,
Des Geschlechts Mund, das da kommt, huldiget dir.

Mother of the Braves! Suevia!
You silent! Aeons rejoice at you,
Men of light you educate beyond number,
The mouths of coming generations will pay homage to you.

It is not difficult to find out who is the other man who could shed light to the modest neighbourhood of Lauffen and Weil. The poet's youthful ambition is reported by his bashful personal letters. Kepler's greatness, however, was revealed to him by an even greater person, Newton. Newton, the pride of Thames, bows his head at Kepler's grave, saying:

»*Du begannst, Suevias Sohn! wo es dem Blick*
Aller Jahrtausende schwindelte;
Und ha! ich vollende, was du begannst,
Denn voran leuchtetest du, Herrlicher!
Im Labyrinth, Strahlen beschwurst du in die Nacht.

You started the work, Son of Suevia!
Where the vision of all previous millennia
Were dazed; but I shall confine what you began,
You shed light before anyone, Superb!
Into the labyrinth, your rays have spellbound the night.

This praise seems to be somewhat parochial. Kepler, as seen in Lauffen, appears to be great because an even greater person (someone from London!) acknowledges him as such. However, the content of the praise shows the poet to be educated and well-informed. He understood that Newton's celestial mechanics, hence the theory of mass attraction, was based on Kepler's empirical laws. This is recognized by him as one of the greatest achievements of human thinking. This makes Kepler's fame eternal. The poem is not just an example of the poet's familiarity with science. It is an honest homage paid to rational thinking, which understands Nature.

Hölderlin's views were shared by many in those years. Newton's authority was unchallengeable; his statements were beyond doubt. Still, why is law of gravitation important to the poet who otherwise glides over the waves as an enamoured swan? It seems that the tradition of Antiquity, which he assimilated

Fig. 9.1 William Blake: Newton

in full, meant something more to him than an archipelago in the middle of a caressing sea, meandering rivers, bold sailors and far-sighted merchants, Athene's people in spirited debates or the heroism against the Persians and the sun shining over the pillars of Acropolis. Hellenism also meant clear thinking, solid logics and well-founded knowledge as he reported in an early letter: "… *returning home from the plane-tree grove of Ilyssos where I spent my time with Plato's disciples and my eyes could follow the majestic flight of the spirit over the dark skies of ancient times…"*. And the hero of his drama was Empedocles the philosopher. The son of Antique rationalism greeted its present-day rebirth. The catchwords are Kepler and Newton. The romantic soul is not their enemy.

Not always the enemy. Blake thinks differently. The picture what we have in several versions is a coloured engraving (Fig. 9.1). It represents a naked youth with fine muscles sitting on a rock, having open compasses in his hand, leaning forward over a roll and constructing a circle in a triangle. The picture is entitled *Newton*.

Looking at this Greek-style young man, one is inclined to consider the perfect body to be the representation of the flawless mind. An idealization which may have been embraced even by Hölderlin. Blake tells us something different.

Hungarian essayist László Földényi wrote an important essay about this picture; what follows now is based on his work. According to Földényi, the menacing black void above the man's head represents the absolute space of Newton's *Principia* in which there lurks the frightful nightmare of God's death. Blake hated and despised Newton, regarding him as one of the most harmful minds of the modern age and the spirit who executes a fatal blow to human imagination. According to Földényi, *"Newton is deprived of heavenly vision and when constructing the world, he can have recourse only to his sense."* He possesses single vision only, which to Blake is the most horrible of losses:

> *Now I a fourfold vision see,*
> *And a fourfold vision is given to me;*
> *'Tis fourfold is my supreme delight*
> *And threefold in soft Beulah's night*
> *And twofold Always. May God us keep*
> *From single vision & Newton's sleep!*

Blake considers double vision to be his own special ability. He sees a thistle in the road:

> *With my inward eye, 'tis an old man grey,*
> *With my outward, a thistle across my way.*

It is obvious that no scientific world view is compatible with such twofold vision. Science does only harm to spirit and mind. A short poem which starts with scolding *Voltaire* and *Rousseau* ends with the lines:

> *The Atoms of Democritus*
> *And Newton's Particles of light*
> *Are sands upon the Red sea shore,*
> *Where Israel's tents do shine so bright.*

The tents of Israel invoke that age when men lived in complete and natural harmony with God. Everything that science can promise either in the Antiquity or nowadays is nothing else but grains of sands rolled over by the waves.

Hölderlin and Blake used the same symbol in their debate (obviously they never heard of each other). Everything that is related to the Greek past is dear to Hölderlin. So are conceptual thinking and even mathematics. This is known also to Blake: Newton, depicted by him as an able-bodied, perfect Greek athlete, holds the compasses which expose Man's soul to bare rationalism. In his painting, the attractive young man embodies Evil. He hates the sunshine glittering over the marble.

Walking in front of the British Library one encounters a huge bronze statue, the sculpted version of Blake's Newton. One feels somewhat bemused—what does it symbolize there?

Bibliography

Földényi LF. Newtons Traum: Blakes Newton. Berlin: Matthes & Seitz; 2004.

Szerb A. Blake. In: Szerb A, editor. Gondolatok a könyvtárban. Budapest: Révai; 1946. in Hungarian.

10

Overcoming the Daze

"*…this gas raised my pulse upwards of twenty strokes, made me dance about the laboratory as a madman, and has kept my spirits in a glow ever since. Is not this a proof of the truth of my theory of respiration? for this gas contains more light in proportion to its oxygen than any other, and I hope will prove a most valuable medicine.*" This glorious report about the effect of laughing gas (dinitrogen oxide, N_2O by its proper name and formula) was written around the end of the eighteenth century by an ambitious twenty-one-year-old youngster, Humphry Davy. Drunkenness, a high spirit and a gas which contains light? Davy's name is associated in the annals of chemistry with important results: he was the first to prepare sodium and potassium, one of the first among those who prepared boron, proved chlorine to be an element and it was he who invented the safety lamp, to be named after him, to mention only some of his achievements.

35 years the junior of Lavoisier, 12 years that of Dalton, Davy lived in an age when chemistry had become a branch of sober science, based on reproducible measurements, requiring calculations, being investigated by well-trained specialists. Davy was a most meritorious protagonist of these years. However, given his nature and inclinations, it was certainly no easy job for him to tame his own mind.

Born into a poor family, orphaned at an early age, fated to labour instead of learn, this was his background. Added to this was a great ambition to self-education, which was both most insatiable and whimsical. The curriculum given below (much abridged) was composed by the young apprentice of surgery and apothecary for himself:

© Springer Nature Switzerland AG 2019
R. Schiller, *Between One Culture*, https://doi.org/10.1007/978-3-030-20538-6_10

1. Theology; 2. Geography; 3. My Profession (i) Botany, (ii) Pharmacology, (iii) Nosology … (vi) Chemistry; 4. Logic; 5. Languages (i) English … (vii) Hebrew;… 10. Mathematics

It is small wonder that the young man, still almost adolescent, being in want of any outer control or advice, was planning to learn everything in a flash, without any system or selection. It is much more surprising that in Davy's time, this type of book devouring and topsy-turvy learning resulted in an intellectual excellence.

Dr. Beddoes, a medical doctor by training, chemist and geologist by inclination, wanted to obtain recognition as a scientist. That was the motive that prompted him to found his Pneumatic Institute—it is likely that the important results of contemporary chemistry, based in a good part on gaseous substances, made fashionable the study of gases. Beddoes promised nothing less than to cure all existing ailments—although, to his credit, he at least treated his patients for free. It was pure scientific interest that motivated him. Davy started experimenting and continued his research at Beddoes's institute and soon he produced a complete book entitled *An Essay on Heat, Light and the Combinations of Light*. A horrible work of the 21-year-old apprentice, which he later retracted himself, but which still contains some excellent observations and surprising ideas. One of them is his theory of burning.

Lavoisier's theory of burning must have been known to young Davy. Lavoisier proved that burning is nothing else but combination with oxygen—at least, that is what we learn at school. It is almost true. Lavoisier regarded oxygen to be some sort of compound consisting of "oxygen base" and "caloric", that is, heat substance,

$$oxygen\ base + caloric = oxygen$$

Burning means that the oxygen base reacts with the inflammable substance, whereas caloric is set free. So it must be because we feel the cosy warmth of the stove.

I think two aspects of nature appear in this idea simultaneously. One of them was most timely: it is the conservation of heat (it is true under certain conditions: no work is done at the expense of heat and the system is thermally isolated). The other one is a hangover from the alchemists: any substance set free during a chemical transformation must be contained by the initial substances. Burning produces warmth, i.e. caloric is set free; hence, caloric must have been a component of the oxygen, which is indispensable for burning.

Since the time of Newton, nobody doubted that light is a material substance. Burning usually goes with light emission. Thus, in parallel with Lavoisier's thoughts, oxygen must contain light substance. Davy considered oxygen to be a compound of oxygen base and light,

$$oxygen\ base + light = oxygen.$$

He tried to support this idea with more or less well-devised experiments. Nevertheless, few years later, he felt ashamed of this work and apologized for his impatience in foolishly generalizing early and undigested knowledge. These were the circumstances that took Davy to his laughing gas experiments.

Chemistry and physiology aside, pneumatic studies also had literary consequences . In those days, the use of narcotics was not kept in secrecy; it was regarded as harmful, but it was not criminalized. De Quincey's work, *Confessions of an English Opium-Eater*, refers openly and unreservedly to the sources and reasons, the unrivalled pleasures and the horrible devastation of opium. More than one contemporary shared this passion. Thus, it was almost natural that Davy's observations soon obtained a fame with important people who developed the habit of visiting the Pneumatic Institute to enjoy laughing gas. So did, for example, two poets, Southey and Coleridge (himself an opium eater), or the members of the rich Wedgwood family. The gas became a desirable luxury and a favourite object of chat in the high society.

The poets Coleridge, Wordsworth and Southey held Davy in high esteem, expressing this in both words and writing. Their friendship, praises and encouraging letters turned out to be a drug to Davy, more dangerous than laughing gas. The ambitious young man considered himself to be an eminent poet as well. Let us be exposed to a short passage of his poetry.

> *Majestic Cliff: Thou birth of unknown time,*
> *Long had the billows beat thee, long the waves*
> *Rush'd o'er thy hollow'd rocks ere life adorn'd*
> *Thy broken surface,…*

The lines go on in the same manner. No doubt, Coleridge and his great friends knew everything about language, poems and poets. Still, one cannot but wonder why Southey wrote to the young man the following lines: "*Your 'Mounts Bay' my dear Davy, disappointed me in its length. I expected more, and wished more, because what there is is good;… I have felt it from the rhythm of Milton… I believe a man who did not understand a word of it would feel pleasure...*" I think it is rather unique that a celebrated poet should reproach the dilettante for the brevity of his writing.

Davy was severely endangered by his swift success. It was a narrow escape for him not to become an attractively mannered scientific conman in high society. Probably, he was rescued by Count Rumford who invited him to join the Royal Institution. In turn, the Institution was rescued by Davy. The founders of this establishment wanted to gather an audience of persons who are both generous with money and have an interest in science. Unfortunately, such people did not show up in great numbers. At least, not until Davy started his lectures on the *Galvanic Phenomena* as it was reported by the *Philosophical Magazine* in 1801. The title, translated to present-day usage, means electro-chemistry, a field which offered eye-catching phenomena to the audience, delivered by a good-looking presenter, who was an excellent communicator, whereas Davy himself could find the central subject of his own research (Fig. 10.1).

The whole story was far from a fairy tale, the ideas of the Managers of the Royal Institution on contemporary chemistry were very different from those of Davy. The first task he was given was to understand *The Art of Tanning*. He

Fig. 10.1 Cartoon by James Gilray on a meeting of the Royal Institution. The bellows are operated by Davy, count Rumford is observing from the right hand side of the picture, the lecturer D. Garnet stops the nose of a poor attendee

set to work with a thoroughness which surely deserved much praise. Having spent months in tanneries in vain, he was unable to raise the attention of London's high society. It was only after this failure that he could deal with the most pressing problems of modern chemistry.

By that time, however, he had come round after the daze of the laughing gas. According to his writings, he understood exactly the task of the researcher together with the necessary intellectual and material means *"In a great country like this it was to be expected that a fund could not long be wanting for pursuing or perfecting any great scientific object. [...] The progression of physical science is much more connected with your prosperity than it is usually imagined."* This type of argumentation is well-known to most of us. As far as I know, Davy obtained the large galvanic source he asked for in this letter. Coleridge, for sure, remained a frequent visitor of his lectures. He said he did it in order to refresh his stock of metaphors.

Bibliography

Collected Works of Sir Humphry Davy (1839, 2001) Vol II Smith, Eder, Cornhill, London, Reprint edition, Thoemmes Press, Bristol.
Hartley H. Humphry Davy. London: Nelson; 1966.
Knight D. Humphry Davy: science and power. Oxford: Blackwell; 1994.
Thorpe TE. Humphry Davy. London: Cassell; 1901.

11

Master Shoemaker and Meistersinger: Two Great Physicists

Analogy—a most powerful tool of human thinking. The happy realization of the resemblance of two objects, two processes or, yes, two ideas. If one of the two is known or understood, we have some hope to master the second one in a similar vein. The two items are not equivalent or interchangeable; their properties are usually of different nature and origin. But there are some common features which might point to a fruitful way of understanding by making use of the similarities even if they are distant and faint. Often we are taught at school through illuminating analogies: a radio wave is similar to a water wave. It is not—but the two are indeed analogous.

William Rowan Hamilton, the great Irish theoretical physicist and mathematician of the first part of the nineteenth century, started as an all-round child prodigy. He was still very young when investigating the mathematical consequences of Newtonian mechanics he revealed that the motion of bodies and the propagation of light can be described in terms of similar mathematical expressions. This analogy has led to most important consequences.

These results were soon acknowledged: he was invited to a university chair and became Astronomer Royal of Ireland. Still, he was not convinced that he was born to be a mere mathematician. Perhaps also poetry? Travelling to the Lake District, he got acquainted at Ambleside with *Wordsworth*. Despite the large difference of age, 35 years, they immediately made friends. On the evening of their first encounter, they walked together for hours and miles between their lodgings before being able to part. The discourse with the great man arose Hamilton's poetical inclination so he wrote a long love poem and sent it to Wordsworth. The answer was immediate, friendly and hard. The poem was dissected, its technical blunders pointed out, and the poor result was excused

© Springer Nature Switzerland AG 2019
R. Schiller, *Between One Culture*, https://doi.org/10.1007/978-3-030-20538-6_11

only by the author's youth. Hamilton answered in the classical style of a dilettante: no better result can be expected from someone who spends only his leisure on versification his life being sacrificed to science. To the credit of both of them, this correspondence did not spoil their friendship, although Wordsworth was far-sighted enough to usher the young man towards mathematics, i.e. to a safe distance from poetry.

This story might discourage anyone from trying to enjoy Hamilton's poetry. Still, let us quote a short passage.

> *Yet 'twas the hour the poet loves*
> *Alone to wander through the groves,*
> *Unheeded, uncontrolled, to pour*
> *His spirit forth in verse to soar*
> *Up to heaven of heavens, to climb*
> *Above the bounds of space and time;*
> *To call ideal worlds to view,*
> *His own creation bright and new.*

At the time when this verse was composed, he also wrote a letter explaining how much he is afraid of being completely immersed into his scientific work without any restraint; literature shall prevent him from becoming nothing more but a mathematician. This is the real essence of the poem, not wandering and groves.

Talking about space and time seems to be nothing more than a poetical common place. Space and time are common place also for Hamilton's pen but with a difference. For him space invokes the science of space: geometry. Whereas time … Hamilton's first work on algebra carried the title: *A preliminary and elementary essay on algebra as the science of pure time.* Locations of things and of events in space are dealt with in geometry; temporal succession and order of operations are treated in algebra. (It is to be noted parenthetically that this metaphysical aspect is based not only on Kant's but also on Coleridge's ideas.)

Thus, the sixth line tells us clearly that the poet-mathematician wants to overpower his own science and is about to create a non-mathematical poetic world. It is completely in vain to ask this genius of theoretical physics about his ideas; he remains firm to keep silent in his poems about the most splendid and really immortal part of his mind. Let us keep science and poetry apart! Wordsworth was right: a person who is ready to mutilate himself for the benefit of some undefined and thin poetic ideal must be an incurable dilettante.

The great analogy between mechanical motion and light propagation was developed further by *Schrödinger* in the early twentieth century. Based on *de Broglie's* idea, he attributed wave-like properties to small bodies, e.g. electrons, and in order to describe their motion he applied the mathematics of the propagation of light waves. This is the basic idea of wave mechanics. I wonder, if it is a mere coincidence that Schrödinger too kept writing poems. He even published them when becoming old. Does analogical thinking go with a leaning towards poetical expression? Analogies are not the sole way of thinking in science; on the contrary, it is rather an exception than a rule. Two other giants of physical theories in the previous century, *Planck* and *Heisenberg*, followed a different path, revealing their theorems by the scrutiny of experimental data. They found mathematical expressions which described the observations of others and investigated the physical content of the mathematics afterwards. This is the opposite of Schrödinger's method. Nothing is known about their poetical activities, but both of them were excellent pianists.

Schrödinger composed his poems as a physicist. The similes in the poem *Parable* seem to have come from the laboratory with the aim to illuminate the age-old and worn-out poetical complaint on "the power of fate". In fact, their object is the physicist's accidental struggle for understanding.

Parabel

Was in unserem leben, freund,
wichtig und bedeutend scheint,
ob es tief zu boden drücke
oder freue und beglücke,
taten, wünsche und gedanken,
glaube mir, nicht mehr bedeuten
als des zeigers zufallschwanken
im Versuch, den wir bereiten
zu ergründen die natur:
sind molekelstösse nur.
Nicht des lichtflecks irres zittern
lässt dich das gesetz erwittern.
Nicht dein jubeln und erbeben
ist der sinn von diesem leben.
Erst der weltgeist, wenn er drangeht,
mag aus tausenden versuchen
schliesslich ein ergebnis buchen. –
Ob das freilich uns noch angeht?

Parable

My friend, what in this life
Weighty and important seems,
Whether causing dark depression
Or gladness and rejoicing,
Deeds, thoughts and wishes
Believe me, means no more
Than a pointer's fluctuations
In an experiment that we design
To fathom Nature:
Merely molecular collisions.
Nor does the light spot's crazy flutter
Let you smell out the basic law.
It's not your joy and trembling
That makes sense of this life.
The World Spirit, if it goes about it
May from a thousand experiments
Enter finally a result –
Is it really any of our doing?

One of the greatest creative minds of physics in the twentieth century maintains that he is supported by transcendental powers in his endeavour to understand the material world. Surprising, if one considers the practical aspects and consequences of the Schrödinger equation. It is perhaps less surprising if one recalls the physicist's active interest in ancient Greek philosophy: here he could recover the unity of rational thinking and emotions, of science, ethics and religion.

Schrödinger, the poet, is deeply rooted in German literature. Although he writes the initials of the nouns with lower-case letters contrary to conventional German orthography, but in accordance with *Stefan George*, his noted contemporary, a part of his poetry has more to do with *Goethe*, than with his own period. This, however, is none of our present business.

These songs are not about science but about scientists.

Bibliography

Moore W. Schrödinger life and thought. Cambridge: Cambridge University Press; 1989, 2015.

O'Donnel S. William Rowan Hamilton, portrait of a prodigy. Dublin: Boole Press; 1983. http://web.mit.edu/redingtn/www/netadv/SP20141215.html

12

Dante Standing on His Head

There is little else he can do at the end of his journey through the Inferno. Progressing towards the centre of the Earth, he finds a deep well with damned Lucifer's vast body frozen in ice. *Dante* and *Virgil*, his master, try to crawl forward grabbing the hair of Satan "*between the matted hair and frozen crust*". To Dante's astonishment, at once Virgil

> *Turned his head round to where his legs had been*
> *And grabbed the hair, like a man climbing up,*
> *So that I thought we'd headed back to hell!*

Unable to understand this change of direction, Lucifer also seems to him to have done a handstand. Virgil explains:

> *"You picture yourself still*
> *On the other side of centre where I caught*
> *The hair of the vile worm that pierced the earth.*
>
> *You were there as long as I climbed downward.*
> *When I turned myself round you passed the point*
> *To which all weight on every side pulls down."*

Satan's huge body traverses the centre of the Earth; thus, if one moves along him, at a certain point, near his hip, everything is seen to turn upside down.

In the time of the Autumn of the Middle Ages (borrowing the title of Huizinga's book), to be more precise on April 9, 1300, Holy Saturday evening, an educated, well-versed person knew that Earth is of a spherical shape and

© Springer Nature Switzerland AG 2019
R. Schiller, *Between One Culture*, https://doi.org/10.1007/978-3-030-20538-6_12

also that free falling bodies move towards its centre, hence the directions up and down have only relative meaning.

This was so in spite of *Aristotle*'s unchallenged authority in matters of physics (as in most other things). He was of a different opinion (*Physics*, Book 8): "*Of things to which the motion is essential some derive their motion from themselves, others from something else: and in some cases their motion is natural, in others violent and unnatural. […] And the motion of things that derive their motion from something else is in some cases natural, in other unnatural: e.g. upward motion of earthy things and downward motion of fire are unnatural.*"

This view was generally held in the Antiquity, as witnessed by Lucretius: "*And yet we do not doubt, I trow, but that all these things, as far as in them lies, are borne downwards through the empty void. Just so, therefore, flames too must be able when squeezed out to press on upwards through the breezes of air, albeit their weights are fighting, as far as in them lies, to drag them downwards.*" Erroneous as this text might be, it sounds in a way natural. As far as the absolute meaning of up and down is concerned, the above two quotations are in accord.

Dante's understanding of nature might well be appreciated; we, however, nurtured by Galilei and Newton find it obvious. Free fall and planetary motion are well known to obey the same law. Newton wrote: "[…] *for a stone projected is by the pressure of its own weight forced out of the rectilinear path, which by the projection alone it should have pursued, and made to describe a curve line in the air; and through that crooked way is at last brought down to the ground; and the greater the velocity is with which it is projected, the farther it goes before it falls to the earth. […] till at last, exceeding the limits of the earth, it should pass quite by without touching it.*"

In the early twentieth century, the interest of several physicists was raised by a well-known phenomenon. It was some 80 years earlier that *Robert Brown* observed pollen grains floating in a liquid perform "*trembling motion*". First, he thought that it is due to living organisms and hoped to have discovered "*the elementary molecule*" of living substances. The phenomenon, however, was seen with grains of inorganic origin as well, so he probably had to attribute it to convection (i.e. to currents within the liquid caused by uneven temperatures).

According to present-day textbooks, the correct theory of the phenomenon is that of *Einstein*. This is almost true. It was certainly he who first assumed that the random walk of the grains, which are large enough to be seen with a common optical microscope, and the diffusion of molecules, which are many orders of magnitude smaller, obey the same laws. Making use of this insight, he developed the statistical physics of diffusion, showing that diffusion is nothing else but the independent random walk of the molecules in the medium controlled by mere chance. Only the velocities of motion are different for a pollen grain and an atom. The word "almost" above refers to the fact that Einstein was not aware of the experiments, perhaps he never even heard of Brown. He just imagined the phenomenon—apparently pretty well.

Jean Perrin, an experimentalist, was perfectly aware of the earlier work. His aim was to prove Einstein's assumption regarding the similarity of Brownian motion and diffusion. He described the microscopic appearance of the floating particles as "*they come and go, stop and start, rise, and sink, and rise again without having the least inclination to stop for good*". Perrin had two great ideas and, curiously enough, it looks as if the two were in contradiction. One of them is in accordance with the above-quoted sentence where zig-zag motion and settling or non-settling is seen as one and the same process. Whereas the other one seems to forget about Newton, Galilei or Dante in the Inferno, since here the main point is the difference between rising and sinking.

This is due to energy conditions. Both microscope and moving particles are exposed to the gravitation of the Earth. It makes no difference to the particles if they go left or right but it influences their rising or sinking. When they rise, their potential energy increases, and when they sink, it decreases. Statistical mechanics, the discipline which deals with the energetics of the molecules in a macroscopic portion of a substance, teaches us that molecules of high energy are always fewer than those of low energy. Now, if the grains which perform Brownian motion are closely similar to molecules, one must find fewer particles to float near the top of the liquid than close to the bottom. Perrin performed a great number of ingenious and exacting measurements finding that molecular statistics is also valid for microscopically visible particles. Einstein's assumption and his model of diffusion were proven by these measurements. Moreover, these measurements offered a direct path to counting the molecules in a gas.

The most important fruit of these investigations has been the undeniable similarity of visible grains and invisible molecules, thus rendering a final, graphic proof for the atomic-molecular structure of matter. Grains of pollen and molecules behave alike. One can even count them.

Twentieth-century modern physics is said to be highly mathematized and graphic models are wanting. This is thought to be a serious difficulty as compared with classical physics. So it might be. This, however, is a consequence of the nature of the problems. Newton was able to see a planet or a pendulum; Heisenberg was unable to see an electron. Still, Perrin's experiment rendered a most visual proof to the existence of molecules, entities which are invisible and belong to the realm of modern physics. Scientific ideas wander along a tortuous footpath.

Bibliography

Aristotle. Physics. Hardie RP, Gaye RK, Translators. http://classics.mit.edu/Aristotle/physics.8.viii.html

Dante Alighieri. The divine comedy. Cotter JF, Translators. http://www.italianstudies.org/comedy/Inferno34.htm

Lucretius. On the nature of things. Cyril Bailey, Translators. http://files.libertyfund.org/files/2242/Lucretius_1496_Bk.pdf

Newton's system of the world. (1846) Published by Adelaide, New York. https://ebooks.adelaide.edu.au/n/newton/isaac/system-of-the-world/

Perrin J. Annales de Chimie et de Physique, 8ème Series, Sept. 1909. Soddy F, Translators. In: Nye MJ, editors. The question of the atom. From the Karlsruhe Congress to the First Solvay Conference, 1860–1911. Los Angeles/San Francisco: Tomash Publishers; 1984.

Brown R. A brief account of microscopical observations…. In: Magie WF, editor. A source book of physics. Cambridge, MA: Harvard University Press; 1963.

Part IV

Novels

13

Chemistry: A High Romance

Chemistry today means science, technique, industry. Chemistry some time ago meant the promise of the Great Epiphany.

The second child out of eleven in a German noble family, *Georg Philipp Friedrich von Hardenberg* was born in 1772. His father being a civil servant, the education of the eldest son followed the obvious path, reading law in Jena and Wittenberg nurturing an already vivid interest in philosophy, history and literature. Having graduated at 22, he joined the county administration. About a year later, he got another job with the central office of the salt mines at Weissenfels. After that, he studied geology, petrography and iron smelting, together with chemistry, mathematics and biology at the Academy of Mining in Freiberg. I do not know if his scientific interest had taken the young man to the mines or, on the contrary, his sense of duty had induced his new studies. Anyway, his interest in chemical theories was genuine and serious.

In those years, ardent fights raged among chemists about the old and new theories of burning. The old and German idea by Stahl maintained that every flammable substance contains phlogiston, burning being nothing else but its escape from the burning substance. The new and French tenet by Lavoisier was that burning is nothing else but union with oxygen. Some years earlier, the fight had already gone beyond the limits of disciplined scientific discussion. Madame Lavoisier burnt Stahl's books publicly in Paris, whereas in Berlin Lavoisier's effigy had been condemned to death. By the time of Hardenberg's studies, this international scientific discussion appeared to be farcical; the court of the French revolution had Lavoisier executed in reality.

Hardenberg was a thorough student attending the lectures of both the pro-phlogiston *Weinreb* and the anti-phlogiston *Lampadius*. His voluminous

© Springer Nature Switzerland AG 2019
R. Schiller, *Between One Culture*, https://doi.org/10.1007/978-3-030-20538-6_13

notes, the so-called *Salinenschriften* (salt mine writings), show the profile of a well-trained, thorough and conscientious expert. He reports his geologic and petrographic observations, problems of coal mining and, obviously, the administrative and business matters of the day. Also, he is engaged in economic and social questions, the horrible labour conditions of the workers or the self-sufficiency of Saxony with mineral resources. But beyond the daily problems of the office, he thoroughly deals with the actual theories of chemistry, this being a lush pasture for the poet.

Yes, because this hard-working technical officer is called *Novalis* by the history of literature and is held to be the great poet and novelist of early German romanticism. One of his best known works is a novel entitled *Heinrich von Ofterdingen*, which remains a fragment due to the author's early death, The blue flower, an important motive of the novel, has become the symbol of the early Romantic Movement. Curiously enough, it appears only sparingly in the book. More recent essays propose to talk rather about the chemistry of romanticism in connection with the novel. The plot taking place in the vague Middle Ages the novel is a *Bildungsroman*, at least on the face of it, since it tells the formation of the protagonist's personality. It is also about travels and adventures; among other trials the characters have to wander in caves and mines, making good use of the author's expertise in mining engineering.

The longest and most important chapter is a tale, a story within the story, told by the wise old man, Klingsohr. (Klingsor, by traditional spelling, the magician of medieval German folklore, who came from Transylvania to Thuringia, prophesied the arrival of Saint Elisabeth from Hungary—at least according to some sources. Novalis used only the name.) The tale is set in a prince's court, the persons being spirits, deities of sundry mythologies, earthly and unearthly beings. The web of symbols is so much entangled that the notes of present-day editions do not try its complete decoding, since poetry, philosophy and scientific observations form a mixture, the parts of which are impossible to separate. Let alone the regularly appearing and decently described erotic pictures of even lesbian and incestuous affaires. Novalis himself speaks about the complicated arabesque of motives. His aim was to construct "*the experimental physics of the spirit*". Let us be limited now to the scientific aspects with which the work is throughout interwoven. These references disclose its age, the age of the birth of modern chemistry, bidding farewell to the alchemist past (Figs. 13.1 and 13.2).

The Queen sits on a high throne chiselled from a sulphur crystal. Sulphur being an important substance for the alchemists, it retained its central role also in a later development: it is one of the three Primes of Paracelsus. The throne, if rubbed, attracts its environment with insuperable force—obviously

Fig. 13.1   Paul Klee. Illustration to Novalis

Fig. 13.2   Paul Klee. Illustration to Novalis

it is electrified. This is Licthenberg's age, who became famous for his electro-static experiments. An iron rod, if whirled around, points to the north—mag-netic phenomena were a favourite field of studies. The rod, however, is bent by one of the persons into the shape of a serpent representing the uroboros, the alchemists' beast which devours itself. Atlas, having been burnt into ashes during the course of the events, must be revived. A silicate mineral called tourmaline is used for this which, again, if rubbed, gets charged and attracts the ashes. Then "*Gold puts a coin into his mouth and the Gardener* (earlier it has been told that he is the Zinc) *puts a bowl under his loin. Fabula touches his eyes and pours the content of a glass phial to his forehead. As the ointment run from the old man's eyes to his mouth and finally into the bowl his muscles and sinews were struck by the lightning of life.*" The description makes clear that the Giant was revived by the current of a galvanic cell, of course, only after the electric circuit was closed by the trickling electrolyte solution. Novalis is known to have dealt with galvanic phenomena.

There are some further similar passages in the text. As it is stated in an essay, there existed a two-way traffic between poetry and chemistry. Whereas romantic thinking influenced the natural philosophy of chemistry, also certain basic chemical concepts, like *mixture, crystallization* and *affinity*, played important roles in the conceptual constructions of romanticism.

Novalis had a curious relationship with the scientific methods of his own age. Once he wrote angrily: "*The followers [of the cult of reason] work ardently to clear nature, earth, human soul and sciences from poetry. […] Their lamp being insolent and mathematically obedient has become their favourite. They were glad because it was easier to break it than let it play in wonderful colours so they called their great task Enlightenment by its name.*" The mean lamp of enlightenment appears also in the novel. It is used by the unpleasant, busy-body, sinister Clerk, whereas its lamp annoys Eros, a child in the cradle. Mind, the Clerk, puts down everything that is said by another character, the respectable Senses, and hands over the full written pages to a Divine Lady, Sophia, the Wisdom who sinks them in pure water. Sometimes the handwriting disappears completely; on some other occasions a part of it is kept. These pages are then attached into a huge book.

Experience, mind, senses, wisdom, their cooperation and criticism, this is the way of scientific thinking and research as seen by the poet. There is no reason to contradict his views in this respect even today.

However, there are some other questions where caution is in order. According to one of his poems, Novalis envisaged an era when the secrets of creation will not be opened by numbers and figures but by a hidden word. He is not happy with the exactitude of science, with all its mathematical descriptions. "*Recently I treat mathematics with contempt.*" he wrote in one of his letters. "*As far as chemistry is concerned the danger is more serious*" because he might be detracted from the literature, "*nevertheless my old inclination toward the Absolute saved me from the whirlpool of empiricism.*" To his mind, chemistry is neither mathematics nor physics which deal with rigid, static forms only. Chemistry is something different, promising dynamics, heterogeneity and diversity, in contrast to mechanics which is seen by the romantic fervour to be uniform and hence inappropriate for the real understanding of nature. The realization of the great spiritual programme is expected by the help of chemistry. "*All the sciences must be made poetic and I hope to talk much with you about this scientific poetry*"—he wrote to one of his friends.

Nowadays, our ideas on mathematics, physics, chemistry are very different from the views of the late eighteenth century. Passions, mythology, poetry have no room in sciences, since those belong to the realm of feelings. To put it in a poetical and obsolete way: heart has nothing to do with mind. However, the borderlines are perhaps not too distinct; let us think of Kekulé who dreamt the benzene structure, of Heisenberg who was educated by Greek philosophers, of Wigner who compared the researcher's passion to love. A four-line poem completes Klongsohr's tale:

> *Gegründet ist das Reich der Ewigkeit,*
> *In Lieb' und Frieden endigt sich der Streit,*
> *Vorüber ging der lange Traum der Schmerzen,*
> *Sophie ist ewig Priesterin der Herzen.*

> Established is Eternity's domain,
> In Love and Gladness melts the strifeful pain;
> The tedious dream of grief returneth never;
> Priestess of hearts Sophia is forever.

One must not forget, Sophie's name means wisdom.

Bibliography

Liedtke R. Das romantische Paradigma der Chemie. Paderborn: Mentis; 2003.

Novalis. Heinrich von Ofterdingen, from the German of Novalis, (Friedrich von Hardenberg). Cambridge: John Owen; 1847.

Vonessen R. Novalis Naturbetrachtungen in den Lehrlingen von Sais. 2006. http://www.philosophia-online.de/mafo/heft2006-2/RVon-N.htm

14

Time and Two Great Men

Every tale is about time, this being the real stuff of any narration. This is sometimes kept in secret by the author, and on other occasions, it is exposed in the most natural way. It is, however, quite rare that it is written on the façade. "*Deep is the well of the past. Should we not call it bottomless?*"—these are the first sentences of the tetralogy *Joseph and his Brothers* by Thomas Mann. The writer does not conceal his intention to make this great epic flow from unfathomable beginnings, moving along the past of the events, towards the storyteller's present. The innumerable tributaries of the main story obeying the same rule are also compelled to flow in the direction of time. Time itself, however, can behave in curious ways. Mostly it elapses uniformly, as one would expect, but in certain cases it becomes obvious that it cannot be let loose from the events.

Let us read, for example, a scene of the second part, one which seems to be not too important for the main story. Young Joseph has been presented with his mother's many-coloured coat by his father and he is vain enough to strut in front of four of his half-brothers. The rest of them guarded their flocks in a distance of several days' walk. The four who had to see the pampered boy indulging in his self-love and being bold enough to attract all their father's love, embarrassed and disgusted, decide to inform those who are absent. Naphtali is ready to perform the task.

"*His instinct to play messenger, his need to report and communicate had been violently aroused from the start—it tugged at his calves till they twitched. Naphtali was obsessed with the notion of space and its divisive nature. He regarded it as his most intimate enemy and himself as the authorized agent for overcoming it—that is, for removing the differences in men's knowledge that distance created. When something*

© Springer Nature Switzerland AG 2019
R. Schiller, *Between One Culture*, https://doi.org/10.1007/978-3-030-20538-6_14

happened at the spot where he was, in his mind he immediately joined it to some faraway place where no one knew of it, which in his eyes was an intolerable, oblivious, vegetative state that he felt compelled to correct by the propulsion of his legs and the nimbleness of his tongue, in order that, if possible, he then might also bring back some still disgracefully unknown bit of news from there to here and thus equalize the sum of human knowledge. In this case, then, it was the locale of his distant brothers that his thoughts—his before any of the others'—had speedily connected with his present location. Thanks to the intolerable consequences of space, they as yet knew nothing, but they had to know, posthaste. In his soul, Naphtali was already running."

Now that's it, one's calves might be tugged with any force, whatever happens *here* and *now* can be known *there* only later, because there is a span between *here* and *there*, a distance which might be overcome by speedy calves, hence sooner or later the message arrives *there*, but this always takes some time. Sure, it can be shorter or longer depending on how nimble the messenger is. If only Naphtali could run infinitely fast!

Thomas Mann's understanding of science looks somewhat curious. Time elapses unevenly in his earlier work, *The Magic Mountain,* as it is controlled by the density of events; he modestly boasts about this effect in a private letter: "*Earlier I could understand only the 'ideal' side of space and time (based on Schopenhauer) and realized their physical relevance without having read Novalis in its full depth let alone Einstein.*" (He thought that Novalis, the eighteenth-century poet, "*had dreamingly preceded*" the theory of relativity.) This letter was written around the time of the origin of the *Tetralogy.* Apparently, he knew little about Einstein's ideas.

Two years prior to this letter Einstein tried, and not for the first time, to explain relativity for the general public. "*Until then* [until the formulation of special relativity] *it had been a tacit assumption that the four dimensional continuum of events can be separated in a most objective way into space and time, in other words that the notion 'now' has an absolute meaning in the world of events. Now, since synchronicity has been realized to be a relative notion, space and time have melted into a unified continuity, in the same way as earlier the three dimensions of space were regarded as unified and continuous.*" And this is so, because "*the law of light propagation makes space and time inseparable.*" Simply speaking *then* and *there* are naturally connected to each other because light does not propagate infinitely fast, similar to Naphtali who is unable to run infinitely fast.

Did Thomas Mann do some reading about things like that? Later the two great men established a personal relationship when Einstein wrote a letter acknowledging the writer's decision not to return into Hitler's Germany.

During the period of their common emigration in America they were neighbours at Princeton for 2 years. Katia, Mrs. Mann, recalls her naïve guest, Einstein who was making intricate political predictions. *"A nice person he was, still a one-sided genius, wasn't he. […] As far as political flair was concerned he had a meagre share."* Thomas Mann himself seems to have kept aloof of Einstein, writing *"we were good, friendly neighbours with the Newton of our age."* I also doubt if Einstein ever was a keen reader of any novel by Mann. He had been keeping *Don Quijote* at his bedside.

Einstein died a few months earlier than Thomas Mann. The necrology on Einstein may be the last paper of the author published during his lifetime in April 19, 1955, in the *Neue Zürcher Zeitung*. *"His great scientific achievements, about which the layman may have only feelings, will be elucidated by experts. What I loved and adored most in his personality and what I will be keeping in high esteem is his moral stance…"* Not a single word about science, nothing about inseparable time and space.

I certainly do not have any philological clue whether nimble Naphtali's fight with space and time may or may not be connected to the notion of space-time in theoretical physics. As a matter of fact, this is not too important. What counts is the artist's sight which pervades the deep well of not only the past but also of nature.

Bibliography

Einstein A. Mein Weltbild. Berlin: Ulstein; 1962.

Mann T. Letter to Käte Hamburger, 1932. In: Thomas Mann: Briefe 1889–1936. Frankfurt am Mein: S. Fischer Verlag; 1961.

Mann K. Meine ungeschriebene Memoiren. Frankfurt am Main: S. FischerVerlag; 1974.

Mann T. Joseph and his brothers. Woods JE, Translator. Everyman's Library; 2005.

15

X-rays and Love

One of the loveliest and most popular stories of chance discoveries is about X-rays. This is perhaps so because everyone has seen some X-ray pictures. The story may make one think that the phenomenon is very simple (how have people not recognized it much earlier?) but at the same time most mysterious (how to peep into me?). The tale about chance starts with the Crookes tube, a more or less evacuated discharge tube which is the ancestor of all present-day fluorescent lamp and cathode ray tubes (older TV screens and monitors). If the tube contains some gas, it gives light. If it is completely evacuated, it stays dark but there flows some sort of radiation within the tube. This was called cathode ray. Its existence has been revealed through the fluorescence of the glass wall and also through the shadow on the fluorescing bulb cast by any metal object which intersects the path of the rays (Fig. 15.1).

Around the end of the nineteenth century, this equipment was an object of intensive research. Great physicists like *Heinrich Hertz*, who discovered radio waves, the noted electrochemist *Hittorf,* the famous engineer *Tesla*, *J.J. Thomson*, the discoverer of the electron… Among other observations, they could show that a permanent magnet deflects the cathode rays. *Philip Lenard* changed the construction by opening the glass housing and closing it again with a thin metal foil; thus he could study the rays outside the tube. Putting a fluorescent screen in the direction of the surmised way of the rays, he observed the screen to light up even if it was at a distance of several centimetres from the tube.

Röntgen wanted to know whether one could observe the rays at even larger distances. Indeed, he could, the fluorescent screen lit up when it was several metres farther. Were these the same rays? Probably not, because he could

© Springer Nature Switzerland AG 2019
R. Schiller, *Between One Culture*, https://doi.org/10.1007/978-3-030-20538-6_15

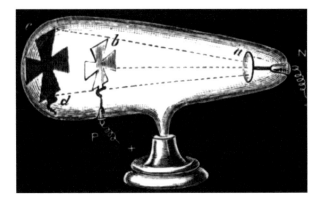

Fig. 15.1 Crookes tube

observe the same phenomenon when he used the original, closed tube with no Lenard-type metal window. In such an arrangement, cathode rays could not reach the screen. The source of the observed radiation must have been the wall of the tube. Next, Röntgen put a photographic plate across the radiation and it was blackened. Were all these observations the outcome of sheer luck? Not exactly. A year earlier, J.J. Thomson observed fluorescence at a large distance from the tube but simply did not care. Crookes realized that photographic plates got dark near the tube, so he ordered his assistant to put the plates to a different place. Several people had seen what Röntgen looked at thoroughly.

During the weeks after the discovery, he did not leave his laboratory where he kept eating and sleeping, even missed his lectures. (Imagine! A German professor!) After two weeks, he produced the first X-ray picture, putting his wife's hand across the rays. The shadow of her bones appeared on the screen. "I saw my own death"—she exclaimed.

After three weeks of intense work, he published his first paper "on a new type of rays" on December 28, 1895, in German and soon also in English. The *Presse,* a Vienna daily, reported about it within a week after the first publication; after another week Röntgen gave a lecture in the presence of the Kaiser (and on the same day the experiment was repeated in Budapest).

Sheer luck? The paper published after three weeks of research already contained all the results, observations and considerations which the state of science of the time could then make possible. Röntgen summarized his observations in 17 theses. Here, together with the large permeability of the radiation, also figured the important fact that magnetic field has no effect on these rays contrary to what was observed with cathode rays. It took 16 years until important new results were here achieved. (X-ray diffraction by Laue, Friedrich and Knipping; Moseley's Law about the frequency of the radiation.)

Röntgen wrote two further papers; after that he changed the subject of his studies.

Medical diagnostics was greatly enhanced by the method. So were also some therapeutic methods, sometimes unfortunately coupled with serious mistakes, though the deleterious effect of the rays was realized soon. Already Röntgen wore a lead apron when performing his experiments. The general public, seeing the X-ray photographs, had the same experience as Mrs. Röntgen: they felt as if they met the skeletons of living people.

The sight of a skeleton had not been alien to European eyes for several centuries. If not earlier than at the time of the devastating Black Death, the bubonic plague in the fourteenth century, the inhabitants of the continent got acquainted with putrefying corpses and bare frames. Exhumations were in regular order in the small graveyards of the towns, the ossuaries at Kutna Hora or Hallstadt with their neat bone and skull stacks can be seen even today. In the *danse macabre* pictures, death appears always in the form of a skeleton, sometimes raising his scythe or holding a sandglass, perhaps playing some merry music, taking king, bishop, ploughman equally to the grave (Fig. 15.2).

But now it is different; the bones of people living with us, moving among us, are openly displayed. They are walking death symbols. X-ray pictures may have recalled several impressions. The connection between Eros and Thanatos, love and death is as old as literature or mythology; the adepts of art and psychology regard this fact as generally understood so it is not fashionable to talk about it any more (Fig. 15.3). Not even about love and the skeleton? This is

Fig. 15.2 From the *Danse Macabre* of Paris, 1424–1425

Fig. 15.3 Hans Baldung Grien, 1517

much ruder. It seems, however, that such rude representations were tolerable already in the past.

However, in Grien's picture, a dead man stands next to a living person. A living skeleton is still something unexpected and after Röntgen's time people had to get used to such conjunctions. This experience turned out not to be alien to gentle feelings.

Earlier a love token might have been a miniature portrait on ivory, a lock of hair, a letter wet with tears, a dry bunch of violets, something like those. Now an X-ray photograph may serve the same purpose. Madame Chauchat takes part with Hans Castorp in a sanatorium for tuberculosis patients around 1910 as described in *The Magic Mountain*, a novel by Thomas Mann. For a patient, the X-ray is not a simple everyday experience but an omen which foretells illness or recovery, death or life. In an ecstatic carnival night Castorp, who during his illness could gather some knowledge in anatomy and physiology, praised, cherished and coddled the woman's body in medical terms. Next morning, just about to leave, the woman presented him with her X-ray picture *"showing not her face, but the delicate bony structure of the upper half of her body, and the organs of the thoracic cavity, surrounded by the pale, ghostlike envelope of flesh. [...] How often had he looked at it, how often pressed it to his lips..."*.

Love cannot be more naked than that. Science seems to give us something more than exact knowledge.

Bibliography

Asmus A. Early history of X Rays, Beam Line. 1995; Summer:10–24, http://www.slac.stanford.edu/pubs/beamline/25/2/25-2-assmus.pdf

Mann T. The magic mountain. Lowe-Porter HT, Translator. New York: A.A. Knopf; 1965

Röntgen WC. On a new kind of rays. Nature. 1896;53:274.

16

Thomas Mann at the Sick Bed

"Disease has nothing refined about it, nothing dignified!" Mr. Settembrini vented his anger on young Hans Castorp as they met in Davos during their regular walk. In those times, in the first years of the twentieth century, Davos was the centre of hoping tuberculosis patients. Settembrini is there due to his illness, whereas Castorp comes to visit his ill nephew, he himself being thought to be healthy. That changes when Hofrat Behrens, the head physician of the sanatorium, detects his tuberculosis infection, earlier in a latent state but flaring up in Davos. *The Magic Mountain*, a great novel by Thomas Mann, covers the seven years Hans Castorp spent at the sanatorium thereafter. Obviously, this is only the frame of the novel and so is the setting of the patients' indolent and idling society. All these serve the author's aim of expressing and depicting his own world view, mental development, philosophy, social ideas and emotions. Castorp, a very mediocre young man, sometimes called *"Life's delicate child"* understands but little out of all that, although the novel seems to represent his evolution, his *Bildung*.

The ambiance of the sanatorium is described in much detail; its inner and outer appearance having been modelled after two really existing ones (Fig. 16.1). The patients' dismal luxury was completed by the reclining chair which made the permanent rest cure comfortable (Fig. 16.2), and by the small vessel called Blue Peter in sanatorium parlance (Fig. 16.3), to collect their sputum in order to avoid contagion.

One portrait, above all, is most authentic, that of the head physician, the merciless but good humoured ruler of the sanatorium, Hofrat Dr. Behrens (Fig. 16.4). His model was much too easy to identify bringing down the wrath

© Springer Nature Switzerland AG 2019
R. Schiller, *Between One Culture*, https://doi.org/10.1007/978-3-030-20538-6_16

Fig. 16.1   Two sanatoriums at Davos which served as models to Berghof

Fig. 16.2   Reclining chair

of the physicians in Davos on the author. Unduly so because the text is not disrespectful of the doctor, it just sees him with the patients' eyes.

Authenticity is based on personal experience. Thomas Mann's wife, Katia who had to stay in Davos on two occasions, spending there several months each time due to some mild tuberculosis infection, was visited by her husband for three weeks, a duration which was the same as the one planned by Castorp before Behrens' X-ray examination finds him ill. Katia was also found ill; nevertheless she died at the venerable age of 97. Decades after her treatment, Dr Virchow, a specialist, questioned the old diagnosis and was allowed to chest X-ray the then 84-year-old widow. He could find no sign of any pulmonary

Fig. 16.3 Blue Peter

Fig. 16.4 Dr. Friedrich Jessen (1865–1935), the model of Hofrat Dr. Behrens

disease. Moreover, he could get hold of her old X-ray photograph kept over decades and during the vicissitudes of emigration: Katia was seen to be completely healthy even there. It seems to have been a misdiagnosis upon which the experience for this great novel rests.

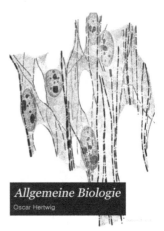

Fig. 16.5 The sources of Thomas Mann's knowledge in biology

Right from the beginning of his stay in the sanatorium, Castorp wants to fit into the society of the patients feeling already a member of it. Illness was regarded by him as a rite of initiation: "*Was he a man who had made his profession on the score of a moist spot, a member of the order, one of those up here...?*" –he indignantly asked. Profession is usually followed by initiation, a process which goes with learning, as the great historian of religion Mircea Eliade writes: "*a person transgresses the natural mode of existence* [...] *and attains the cultural mode by initiation.* [...] *He who is to be initiated becomes worthy of the saintly doctrine only if prepared intellectually.*" So Castorp sets himself to learning. He reads biology, physiology and medicine, since these disciplines may make him understand his own illness. Thomas Mann also had to learn a portion of them in order to describe his hero's intellectual foray. We happen to know the books he used, the works by L. Hermann, J. von Uexküll and O. Hertvig having been acknowledged textbooks of that age (Fig. 16.5).

It is somewhat surprising that Castorp tries to acquire some knowledge also in physics, astronomy and cosmology. The surprise is twofold. On the one hand, these things have nothing to do with his illness. On the other hand, the author himself had little if any interest in science. On the contrary! He had always made a sharp distinction between *humaniora*: the realm of literature, art, philosophy and history, which were important for human thinking and behaviour, and *realia*, dealing with the description and understanding of inanimate nature. No doubt, his world was that of humaniora. It is, however, learning and scientific understanding that take the sick man, who looks for companions in his suffering, into the depth of his malady, into the midst of the society of the sick. Mr. Settembrini tries to save him from this dangerous initiation.

Because he, Settembrini, is a most authoritative and authentic representative of humaniora, advocating the ideas of progress with nineteenth-century fervour. The ideas based on his freemasonic, socially committed, international ways of thinking and at the same time (at the same time!) his fervent Italian nationalism are contrasted with the morally intolerant and conservative views of Naphta, a Jew who had converted to become a Jesuit. Their incessant arguments make up a good part of the novel. (Naphta's physical appearance and manner of speaking was modelled after Hungarian philosopher Georg Lukács.) Their long discussions were held throughout on the territory of humaniora with Hans Castorp understanding but little of the whole business. Thomas Mann looks with marked irony at both of the stubborn disputants without leaving any doubt of his sympathy for Settembrini.

Castorp gathered curious scientific views, things like that the following: "*the atom was nowhere near large enough even to be spoken of as extraordinarily small. It was so small, […] to think of it as material, but rather as mean and border-line between material and immaterial.*" The textbooks of quantum mechanics on the shelf shudder on hearing this. Nowadays, that is. But ten to twenty years before the events of the novel, some great scientists had similar thoughts. One of them was Wilhelm Ostwald who thought that the "theory of energetics" (i.e. well-founded thermodynamics) refutes materialism, since its basic notion is not mass but energy. There was also the great experimentalist Crookes, who, by considering his own cathode ray tube, went as far as Spiritualism. Castorp's views were not too bizarre for his time. I do not know the sources of the author's scientific knowledge. Perhaps, he talked to his brother-in-law, Peter Pringsheim, a theoretical physicist? Anyway, scientific thinking had never touched him; this area remained alien to his devotion to humaniora.

His other great novel connected to illness and a sick man, *Doctor Faustus*, compels science to an even more frightful role. In the nineteenth century, tuberculosis was seen as the slowly devastating, mild malady becoming poets, musicians, tender souls. Syphilis, however, was regarded as a horrible punishment of sexual debauchery, some sort of a worldly foretaste of damnation with all its frightful symptoms and sad end. At the beginning of the novel, in the early twentieth century, protagonist *Adrian Leverkühn*, the great composer, makes a pact with the devil in order to be able to perfect his genius. The price of the contract is syphilis, whereas damnation is nothing else but paralysis making his life end with humiliating suffering. The title of the novel invokes the medieval Faust legend immortalized by Goethe. Some motives

Fig. 16.6 Osmotic garden: water glass + salts of Ca, Fe(III)…

remind the reader of the life of the great German philosopher and poet, Nietzsche.

Initiation again. Leverkühn must be initiated into Satan's practices. Thomas Mann is careful to maintain a medieval atmosphere in the contemporary story by allusions, phrases and an artful style. Again, the task of the person to be initiated is learning. So he does, studying musicology, this being his future craft, theology in order to get acquainted with the machinations of God's and man's enemy, and science in order to please Evil.

At the beginning, it is about harmless natural phenomena, things which the composer saw in his childhood. His father made an *osmotic garden* from water glass and some metal salts (Fig. 16.6). What the child saw was this: "[…] *there strove upwards a grotesque little landscape of variously coloured growths: a confused vegetation of blue, green, and brown shoots which reminded one of alga:, mushrooms, attached polyps, also moss, then mussels, fruit pods, little trees or twigs from trees, here and there of limbs.* […] *"And even so they are dead,"* said Jonathan,* [the father] *and tears came in his eyes, while Adrian,*[the son] *as of course I saw, was shaken with suppressed laughter."* This seeming transition between animate and inanimate moves the heart of the innocent, whereas it is cruelly ridiculous to the one who is chosen to damnation, perhaps just because it deceives the pious observer.

Living organisms are also ready to play such devilish tricks. An innocent-looking butterfly, Hetaera esmeralda by its name (Fig. 16.7), mimics a flower petal in order to deceive its menacing enemy, "[…] *when she flew she was like a petal blown by the wind. Then there was the leaf butterfly,* […] *this clever*

Fig. 16.7   Hetaera esmeralda

creature disappears by adaptation so entirely that the hungriest enemy cannot make it out." The vagueness between the vegetal and the animal appears here as devilishly frightful.

When both body and soul of this great musician are in the devil's grasp, it is only then that he gives himself to science, at a time when he, opposing the centuries-old cosy and moving traditions of music, composes coldly agonizing pieces. (Is this a modernized version of Goethe's Faust? He says: *"So I've given myself to Magic art, /To see if, through Spirit powers and lips, /I might have all secrets at my fingertips"*). Among other subjects, he talks about the structure of the Universe, including its expansion (a mild anachronism since at the time of the discussion in the novel this idea was still unknown). He talks about the extension of the Cosmos, dimensions which are incomprehensible to our minds trained on earthly distances, amazing his meek friend who gets frightened. *"The data of the cosmic creation are nothing but a deafening bombardment of our intelligence with figures furnished with a comet's tail of a couple of dozen ciphers, [...] There is in all this monstrousness nothing that could appeal to the likes of me as goodness, beauty, greatness; [...]"* Most probably the author is also both amazed and frightened.

So are we, not because of the Cosmos but of the author. Does science mean figures and mere figures to Thomas Mann? Is science nothing else but a deafening bombardment of human *intelligence*? Osmotic garden and cosmology, are both the creations of Satan? One might feel it somewhat absurd and not only in our present time with its scientific leaning but also in the often evoked Middle Ages.

Medieval teaching and education rested on the Seven Liberal Arts, which were divided into two groups. Whereas the introductory *trivium* consisted of grammar, dialectic and rhetoric, it was followed by the *quadrivium* with the subjects of astronomy, arithmetic, geometry and music. Thus, at least three of the seven arts dealt with exact, mathematically formulated description of the world. Quadrivium was also a territory in the devil's realm? That would be difficult to believe.

Admittedly, only a single strain of the two great novels was considered now, the author's uncomprehending animosity against exact science. Of course, these books offer uncountable marvels to their readers, whenever one re-reads them one finds some new gems still unearthed. But this sole strain! How did it happen that this exceptional mind, open to synthetizing a treasure trove of intellectual achievements, excluded one of the most fascinating mountain ridges of the human spirit from the notions of humanity and humanism?

Still, we must not lose hope completely!

Bibliography

Bellwinkel HW. Naturwissenschaftliche Themen im Werk von Thomas Mann. Naturwissenschaftliche Rundschau. 1992;45:S.174–83.

Eliade M. Patterns of initiation. Maskell Lectures. University of Chicago; 1956.

Genz H, Fischer EP. Was Professor Kuckuck noch nicht wußte. Reinbek bei Hamburg: Rowohlt Taschenbuch Verlag; 2004.

Goethe JW. Faust. Kline AS, Translator. https://www.poetryintranslation.com/PITBR/German/Fausthome.php; Mann T. Doctor Faustus. Lowe-Porter HT, Translator. New York: Alfred A. Knopf; 1948

Mann T. The magic mountain. Lowe-Porter HT, Translator. New York: A.A. Knopf; 1965

Sontag S. Illness as a metaphor. New York: Farrar, Strauss & Giroux; 1978.

Tuberkulose und Kunst. http://blauerheinrich.jimdo.com/tb-und-kunst/ (n.d.).

Virchow C. Medizinhistorisches um den "Zauberberg" "Das gläserne Angebinde" und ein pneumologisches Nachspiel, Gastvortrag an der Universität Augsburg am 1992 Juni 22; Augsburg; 1995.

17

Augury: Franz Werfel

F.W. (AKA Franz Werfel), a Prague born Austrian writer, living in California during his final years, reincarnated some hundred thousand years after his death. This was just a sort of conventional materialization, the outcome of the usual conjuration of a spirit. Having been buried in his creased tailcoat decorated with his one and only award, his own resuscitated self and also the resuscitating distant posterity must have enjoyed his unchanged looks. After his long stay in the grave, he was surprised by the great changes in the landscape. No clouds in the sky, a dim Sun, lowlands instead of mountains, grey grass with an unknown touch and … "*where are the birds*" asked awe-stricken F.W. under the ominously silent skies.

The birds could not survive that day, 13 November, sometime in the very remote past, when a flash of cosmic catastrophe changed the usual way of earthly nature; or so it is explained in the novel *Star of the Unborn*. The moment when, among other abrupt changes, "*the nitrogen content of the air increased threefold in several seconds*" thus it was a matter of sheer luck that the atmosphere was not set in fire. (Was that the level of chemistry teaching at a secondary school in Prague?) This much chemistry is sufficient to judge the scientific foundations of Werfel's utopia. The book deals, at some point, also with biology and astronomy, unfortunately so. It is spiced with timely political hints, but the author, being a religious man, finds church and belief more important than material experience. However, the birds are greatly missed in that dumb and barren scene.

"*Similar to the birds lifted high by their feathers [...] the soul is lifted in the body by thoughts*"—I found this simile by *Hildegard von Bingen* in an encyclopaedia. The bird, traditionally symbolizing and embodying light,

© Springer Nature Switzerland AG 2019
R. Schiller, *Between One Culture*, https://doi.org/10.1007/978-3-030-20538-6_17

spirit and Heaven's gift, is a serious loss to the comfort-loving future of the novel.

Here is another quotation: "*The sedge has withered from the lake, /And no birds sing.*" These lines by Keats are the motto of the book *Silent Spring* by *Rachel Carson*, published a mere fifteen years after Werfel's novel. Its prophecy ranges only decades instead of hundreds of millennia. The biologist-journalist author calls her readers' attention to the collateral dangers of the use of chemical pesticides which threaten all hot-blooded animals, including ourselves. The main target of her misgiving is DDT which decomposes under natural conditions only slowly. It seems to be easy for her to find a host of biological, medical and environmentalist arguments. Still, the title and the most impassioned, hence most cited, chapter of the book are connected to the loss of birds. I doubt that the author had Werfel's book in mind. She knew, however, only too well that the readers' souls are more easily moved by the termination of the bird song than by any discourse on physiology.

Carson's book has become the manifesto of the fight against the chemical and technical manipulation of nature. Its impact was compared to the role of *Uncle Tom's Cabin* as an appreciation by one of its reviewers. Similarly to Harriet Beecher Stowe's novel which preceded the American Civil War so Carson's book … I wonder if the comparison with the sentimental novel of honest intentions is most flattering to the book which heralded the start of the green movement. Its success, however, is certainly beyond doubt. To mention just one of its achievements: DDT, together with a number of further POPs (*p*ersistent *o*rganic *p*ollutants), has been on the list of banned chemicals for many years.

Otherwise, DDT is a most useful agent being very effective in killing flies which are the vectors of malaria; thus the lives of several million children and adults were saved by its use. But its adverse effects include not just the disappearance of robins and starlings from the gardens of Hinsdale, Ill. It accumulates in the human body causing fertility problems, being carcinogenic and effecting mental retardation. This Janus face is a characteristic not only of DDT but a good number of other chemicals. Hence, balanced compromises were needed and have been elaborated regarding use, limitation and ban of the substances. Birds seem to be saved.

Thus, Werfel turned out to be not too successful in predicting the fate of the birds. He, however, made another prophecy regarding the Labourer. Waking up from his mortal sleep, he could find little if any traces of productive work in the society of the future. Nevertheless, humanity, the population of which has been much diminished (!), lives in unprecedented comfort and welfare. This has nothing to do with technical development; on the contrary, people of the future maintain a supernatural contact with nature. For example, their mode of transport differs radically from the ones we know: we move towards our destination, they move the destination towards themselves. No design or blueprint was attached; still, one serious problem was realized by the writer: chaos would break out when more than one person invites the same destination. This is averted by the vigilance of the coordinators, a bunch of elderly people sleeping like a log.

Material goods, from baby shoes to wedding cakes, from houses to pillows, are produced from sunshine and starlight by the Labourer as he walks easily in a spacious garden. As the only proletarian in a pampered society, it is he who, accompanied by his large family, performs the task that we call production and manual labour.

Let us now forget the starlight, although Werfel diligently explains that similarly to the radio waves which carry information cosmic waves could carry material. All right, let us be generous enough to skip these ideas but let us cast a glance at the number of manual labourers. Werfel may have been a better prophet in this matter. I looked up an old statistical forecast for the first decade of our century, made some ten years earlier, to see what was thought the Americans would live on. Certainly not on heavy labour: the number of farmers was expected to decrease by 25%, that of steel workers by 22% and that of miners by 14% between the years 2000 and 2010. And now some actual figures: in 2004, the total employment in manufacturing was more than 14 million, which decreased to about 12 million by 2010. Probably, the share of the manual labourers in this shrinking was even higher. If one extrapolates these figures to a hundred thousand years, one finds not a single labourer, irrespective of how populous our race would be then.

Augury, prophecy or sarcastic utopia, one thing seems to be certain. An intellect with sharp ears will hear if something is in the air, be it bird song or anything else.

Bibliography

Biedermann H. Knaurs Lexikon der Symbole. München: Droemer Knaur; 1989.

Carson R. Silent spring. Boston, MA: Houghton Mifflin Co; 1962.

Lurker M. Wörterbuch der Symbolik. Stuttgart: Kröner; 1986.

Werfel F. Stern der Ungeborenen: Ein Reiseroman Stockholm. Sweden: Bermann-Fischer; 1946.

Werfel F. Star of the unborn. G. O. Arlt, Translator. New York: The Viking Press; 1946. http://www.sf-encyclopedia.com/entry/werfel_franz

http://www.malaria.org/DDTpage.html
https://www.oecd-ilibrary.org/industry-and-services/total-employment-in-manufac-
turing-2012-1_manu-emp-table-2012-1-en
http://www.worldlife.org/toxics/progreas/pop/ddt.htm

18

Oil of Vitriol in the Marital Bed

This is no family thriller, what comes in the forthcoming passages is a lesson in chemistry. At least that must have been the intention of the author, novelist *Anita Nair. Ladies Coupé,* somewhere between a novel and a string of short stories, is set in India. In earlier times, special coupés were reserved in the carriages of the Indian railways for ladies travelling several days. Six women in such a coupé, spending together the time of a long journey, talk about their lives. It seems that the fate of an Indian woman is neither easy nor pleasant, even if she is well-to-do and has been born in a higher caste, not even around the end of the twentieth century.

Is this just the feeling of an old man, myself, somewhere in Central Europe? I guess not, this is how the young Indian author, having a European cultural background, living in Bangalore, writing in English, also feels. But what is the opinion of the stories' protagonists who are not educated in English literature and European customs? Do we hear the voices of those who tell their life stories or is it the narrator who speaks? Do their experiences and views coincide? Who complains, the narrator or her characters? One has to break down the text into its components and, given the task of separation, chemistry might help.

One of the women teaches chemistry at a secondary school. A talented and hard-working young girl, her family and teachers believed in her brilliant scientific career. She, however, falls in love, gets married and becomes a teacher at a school where her young husband is the headmaster. Her happiness does not last for long; the wife is unable to respect her husband, due to his vanity and greed for power. Neither can she nurture a high opinion about her colleagues at school. She wants to break loose although it was she who made the

© Springer Nature Switzerland AG 2019
R. Schiller, *Between One Culture*, https://doi.org/10.1007/978-3-030-20538-6_18

choice. The author thinks that a chemistry teacher can relate all these only in chemical terms.

"Every time I meet someone, after a few minutes, they cease to be a person. To me that person becomes a chemical. [...] To me, my chemicals were everything." The acquaintances she disliked were characterized in that vein. A red-haired colleague with *"a disagreeable body odour"* was thought to be identical with bromine, the reddish-brown, stinking element. *"[...] maximum precaution had to be taken when handling her."* Another lady is *"light and silvery like the element lithium, she dazzled everyone with her charm and smile. [...] Then there was [...] the Hindi master. There was only one element that could be him— cobalt. Goblin."* This last one is nothing more than the textbook etymology of the name related to old metallurgist experience. Another person is arsenic because he is a *"poisoner of minds"*, someone else, a joker is seen as *"laughing gas"*, whereas a person who easily changes his mind following prevailing conditions is identified with *tetrasulfur-tetranide*, a compound which changes its colour with changing temperature.

How about the husband? *"Biting. Scathing. Colourless. Oily. Dense. Sour. Explosive. Given to extremes. Capable of wiping out all that was water, fluid and alive. Fortified to char almost anything that was organic. [...] Concentrated sulphuric acid. King of chemicals. Oil of vitriol."* The description is technically perfect. Scientists, however, are usually contented with the fact that the sober observations of science are free from emotions. So we are unhappy to see that a compound is hated that much as a spouse. The simile here has been reversed: not the husband is as reactive as the oil of vitriol, but vitriol is as hideous and hateful as the husband. Poor sulphuric acid!

In order to characterize herself, the wife finds chemical expressions which are both contradictory and in no small measure boastful. *"I classify myself as water. [...]it is water in its various forms that configures the earth, atmosphere, sky, mountains, gods and men ..."* But in her unhappy marriage *"I was frozen in a solid state."* Later, however, *"[s]omething happened. A chemical change. There is a technical name for the water that I turned into. Supercritical water. Capable of dissolving just about anything which as mere water, it wouldn't even dare to aspire to. Raging with a vehemence that could burn and destroy poisons that if allowed to remain, would kill all that was natural and good.* "Whereas at a later stage of the story she writes of herself *"But that night I was Aqua Regia. Royal water. All acid and hate."*

I am not going to be over-scrupulous, it is not a technical text we are reading. Otherwise the author is well informed, as far as the properties of supercritical water are concerned, even if the text is somewhat melodramatic. The trouble lies somewhere else. Elements and compounds as described do exist.

Chemists as described do not. That is the reason why I think that the much quoted chemistry is a good reagent for the author's authenticity.

This short story revives the centuries-old caricature of the scientist, handed down in anecdotes and made eternal in literature here, however, without the sarcasm of the original pictures. The blame that cheap and illiterate derision could fasten onto the scientist appears in Nair's text as an intellectual virtue. Things like frogs leaping from the biologist's pocket, a human leg in the anatomist's hand as he is buying a movie ticket, a function expressed in polar coordinates written in the love letter of the mathematician, whereas as far as the chemist is concerned… well he/she is like the above description. Undigested information, pushing technical terms, textbook chapters instead of human relationships, data and rules displacing thoughts and feelings.

No harm is done to the chemist by this picture. No chemist has ever lived married life on the reagent shelf. However, the author's rigid ignorance is telling. No scientist similar to the protagonist has ever been seen. Such a person can only be imagined with a schoolchild's limited phantasy. Now the trouble is this: if the author is as poor as that when inventing a chemist, what shall one think about her invention of the old Brahmin's wife?

Bibliography

Anita Nair A. Ladies coupe. London: Chatto & Windus (Random House); 2002.

Part V

Thermodynamics

19

Thermodynamics of Happiness and Love

It is easy to express happiness in terms of thermodynamics. Let G denote the extent of happiness, E the quantity of energy which is used by our organism in accordance with our own will and, in contrast, W the quantity of energy we are compelled to use against our intention. Hence, $(E + W)$ gives the total energy consumption; the sign of $(E - W)$ shows whether we make use of the energy in accordance with or contrary to our desires; the absolute value of the energy difference gives the extent of our activity be it for or against our will.

A happy person has abundant energy which he/she can use according to his/her resolve and inclination. We can state all this by an algebraic expression:

$$G = (E + W)(E - W) = E^2 - W^2$$

Simple as that. This was stated by the great scientist, Wilhelm Ostwald, one of the fathers of physical chemistry. He knew very well that happiness cannot be quantified; only a tendency is expressed by the law, together with the relative extent of happiness.

A conscientious researcher, Ostwald, analysed his formula investigating which of the quantities can be measured. Happiness, as mentioned above, cannot. But the total energy consumption $(E + W)$ is measurable; it is given by the rate of metabolism; hence, it is proportional to the rate of the body's carbon dioxide production. The amount of voluntary energy, E, is proportional to the energy consumption of the brain—"*this seems to be beyond doubt*". That is, the quantities on the right-hand side of the equation can be subjected to measurements, at least as a thought experiment.

© Springer Nature Switzerland AG 2019
R. Schiller, *Between One Culture*, https://doi.org/10.1007/978-3-030-20538-6_19

We are happy if G is positive; the larger G is the happier we are. Similarly, we are unhappy if G is negative, the more negative, the unhappier. Whenever more energy is expended at our will than against it, we feel happy. In the opposite case, if more energy is consumed in processes which are against our will than in those which are in accord with it, we become unhappy.

This is still not the end of the story. On the one hand, it is a general experience that happiness (its feeling? or extent?) often changes with time, although the external conditions remain unchanged and also the individual differences between people are remarkable. All these seem to contradict the attractively simple expression written above. On the other hand, it is well known that the energy scale lacks any absolute zero point. Fine! This is the answer to the first problem—the energy zero is different for each individual and also might change for any person with time. If one is exposed to certain successes or failures systematically, these events disappear from the mind "*and cease to be the object of any sensation of will*" (*Willensempfindung* is written in the original German text). The zero point of the voluntary energy is different for each and every individual at each and every period of life. Moreover, the zero points of E and W are different; otherwise the value of the zero point would be cancelled in the difference of energies.

The following part of Ostwald's article consists in the definition of behavioural types in terms of the happiness equation together with some advice on happy life. And an important sentence: "*That is why sleep is never accompanied by any feeling of happiness since in that state the energy exchange is exclusively physiological, which is unconscious.*" The date of the article might be interesting: it was written in 1904.

Well, sure, Freud published his *Interpretation of Dreams* four years earlier. In this book very different ideas are expounded on sleep, dreaming and happiness. The world of the unconscious had been revealed here for the first time, a discovery that greatly influenced the intellectual life of the century. Still, it would be unfair to mock Ostwald. For several years, Freud's work was unknown even among experts; initially his lectures were frequented by three students only. In these years, Ostwald's fame was at its height with three years later obtaining the Nobel prize. Having understood chemical thermodynamics in unprecedented depth, he stopped with his chemical research and developed a philosophical system, called energetics, which was based on the two Laws of Thermodynamics regarded by him as the most general principles of nature.

Thermodynamics greatly influenced the thinking of the age. Ludwig Boltzmann was one of the founders of statistical physics and on that account (another) father of physical chemistry. In 1905, he gave a lecture in the

Philosophical Society of Vienna with the title: *"Understanding the entropy law and love by the principles of probability calculus"*. Its text was not handed down to us. However, a contemporary of ours, a Viennese professor of physical chemistry and biographer of Boltzmann, Engelbert Broda, was bold enough to venture a joke and tried to "reconstruct", that is to concoct the text. He made use of several passages of Boltzmann's writings for the general public, and also some motifs of his way of thinking.

The basic idea is simple. The primary energy source of life is the Sun which is abundant enough to maintain life on the Earth. The difference between the radiation of the Sun and the living organisms lies not with their energy content (energy is conserved, anyway) but with their organization. Living structures are highly organized, whereas thermal radiation is completely disordered. *"Thus, the universal fight of living beings is not for basic substances, not even for energy which is present in every being in excess, albeit unfortunately in a non-convertible form. The fight is for the entropy which is to our disposal due to the energy transition between the hot Sun and the cold Earth."*

These are Boltzmann's genuine words—slightly corrected in view of some later development by Schrödinger and Brillouin. Indeed, the entropy of an organized system is lower than that of a disordered one.

Boltzmann, who held Darwin in high esteem, expressed the ideas of struggle for life and natural selection in terms of statistical physics. The race that survives is the one which is more successful in decreasing its own entropy. *"The striving to make the entropy decrease motivated the first protozoan to divide, the cat in heat, or Dante in a dream on Beatrice."* This last sentence is by Broda. But what comes next is a genuine Boltzmann text again. *"That way we understand that it is both useful and necessary for our race to find certain sensory perceptions pleasant, so we try to come across with them, and others offensive. Hence the origin of notions like beauty and truth can be explained on a mechanical basis. The enthusiasm of a young man for the poems of Schiller can be reduced to mechanical causes."* (Boltzmann himself was an ardent reader of Schiller's early works.)

It would be both stupid and unfair to deride these great men. Still, we could politely shake our heads. A taste of doubt about the far-reaching validity of their own ideas might have done some good to their thinking in psychology.

Bibliography

Broda E. Erklärung des Entropiesatzes und der Liebe aus den Prinzipien der Wahrscheinlichkeitsrechnung. Physikalische Blätter. 1976;32:337.

Ostwald W (1910) Theorie des Glücks, In: Forderung des Tages, Akademische Verlagsgesellschaft, Leipzig.

20

Thermodynamics on Stage: Schnitzler and Onsager

The play by Schnitzler was not to be performed in Vienna around 1900 when it was written. *La Ronde* had its first performance in Budapest 1912 provoking a huge scandal even here. The translator thought it wiser to conceal his name; most probably it was Sándor Bródy, himself a successful dramatist of the time. Schnitzler was pretty impolite with the Viennese bourgeoisie, a group of society which prized itself on its high moral standards. Classes might have been sharply divided according to wealth, origin, culture and behaviour; this multifarious society still could be entertained by Strauss's and Lehár's uncompromising mirth. However, no matter how much the houses of the boasting nouveau rich along the Ring tried to compete with the Baroque palaces of the aristocracy, no matter how far the self-esteem of the Monarchy was that of a rich and mighty empire—literature still knew better: the menacing End was nearing.

The contemporaries of Schnitzler regarded him as the most Austrian of all authors, approaching the problems of his world only in terms of psychology. He thought that the superseded norms of behaviour, the lies of society must be demolished at the root of human nature. This most outspoken play expresses the deteriorating, unsafe common feelings in terms of love and sexuality. The content of the play was parodied after the opening night by the Hungarian writer, Karinthy, in the following way: "Fallen Woman loves Soldier, Soldier loves Parlour Maid, Parlour Maid loves Young Gentleman, Young Gentleman loves Young Wife, Young Wife loves his Husband, Husband loves Little Miss, Little Miss loves Poet, Poet loves Actress, Actress loves Count and, in the end, Count loves Fallen Woman—all these make a cycle". (Fallen Woman! Oh bye-gone times! The word "whore" was still not tolerated in print or on stage.)

© Springer Nature Switzerland AG 2019
R. Schiller, *Between One Culture*, https://doi.org/10.1007/978-3-030-20538-6_20

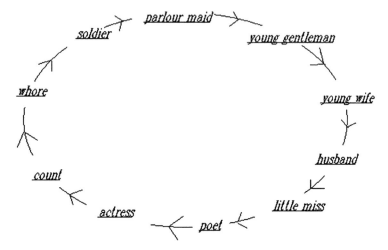

Fig. 20.1 The scheme of the play

A circle of unrequited loves. What scandalized the audience was the fact that finally each couple sleeps together. Ten dialogues about unrequited love, but when words have come to an end the quick curtain has the important task of hiding what is to follow. The simple and melancholy cycle of feelings can be represented by the diagram below (Fig. 20.1).

Several years ago, the Burgtheater in Vienna presented a new performance of the play. Present-day audiences are, of course, difficult to shock by sexuality on stage; words like "indecent" or "lewd" can be found only in antiquated dictionaries. The director was wise enough to resort to grotesque irony: the partners harass each other with rolling iron beds and none of the actors tries to compete with the style of a sex magazine. Now, a hundred years after the first performance, it is easy for us to understand the dismal message of the play about society being out of joint.

Nevertheless, this can be established by casting a glance at the graphical scheme below. Equilibrium is a scientific notion of thermodynamics; this discipline teaches us about the ominous social message of Schnitzler's play. An example may help. Lars Onsager investigated a simple chemical cycle: compound **A** transforms into compound **B**, **B** into **C**, finally **C** into **A**. Using paper and pencil, one can devise two distinct ways along which the concentrations of the compounds become constant in time and the reactions attain a steady state. The two possibilities are depicted below (Fig. 20.2).

One of the ways is denoted by a question mark: here all the reactions proceed continuously and the rates of formations and decompositions are such that each reactant is created as fast as it is decomposed. This diagram is the counterpart of that of the play.

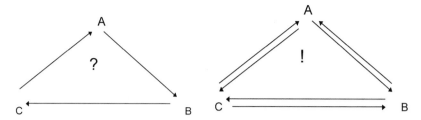

Fig. 20.2 Onsager's cycle

The other one is denoted by an exclamation mark. Here, each reactant maintains a separate equilibrium with the one from which it is formed, and with the other one, into which it decomposes. This is the *principle of detailed balance*. According to both experience and the basic ideas of thermodynamics, this scheme is preferred by nature; stability can be established only in this way. The diagram with the question mark cannot render any steady state.

(Thermodynamically speaking, this result is easy to visualize. The cycle with the question mark is self-contradictory—it assumes an equilibrium where the reactants still form all the time. Such an equilibrium is absurd.)

(Otherwise the principle of detailed balance is an important starting point in science. The celebrated reciprocity relations of Onsager were visualized through the exclamation mark scheme. In that case, the principle means that the transformation rate of **A** is influenced by the concentration of **B** in the same manner and to the same extent as the transformation rate of **B** is influenced by the concentration of **A**. This has nothing to do with Schnitzler, of course.)

Nevertheless, the thermodynamic analogy is tempting. A social-emotional cycle revealed by the play according to its not-much-simplified scheme cannot result in a state of rest. Under such conditions, no steady state is possible. Schnitzler and Onsager knew this pretty well.

Bibliography

http://depts.washington.edu/vienna/documents/Schnitzler/Schnitzler_la_ronde.htm

Onsager L. Reciprocal relations in irreversible processes I. Phys. Rev. 1931;37:405.

21

The Physicist and the Poets

The great Austrian theoretical physicist, *Ludwig Boltzmann*, committed suicide at the Adriatic resort, Duino. Being a well-known figure also beyond the immediate circles of scientists, his death created some stir in the newspapers of 1906 in both Vienna and Budapest.

One of the articles came to an astonishing conclusion about the dangers of mathematics. A young Hungarian poet, Kosztolányi, having attended the series of Boltzmann's popular lectures remembered how fiercely he had attacked Schopenhauer. He did it in a caustic, venomous, satirical style with the scientist's superiority. As far as the poet could remember, the physicist was most upset by Schopenhauer's false views on colourful, glittering, blessed life.

The poet's memory was somewhat imprecise. A year before his death, Boltzmann wrote a short essay about all what he refused in Schopenhauer's ideas. The physicist, first stupefied "*by the mode of expression associated in the past with fishwives*", expounded "*Schopenhauer was not at all felicitous in what he called a priori*". For example, Schopenhauer stated the three dimensions of space to be a priori self-evident. Nowadays (in 1905), physicists find easy to imagine a space with more than three dimensions or even a non-Euclidean space. Obviously, the question is not whether our space is Euclidean or not; one must, however, understand what "a priori evident" is and what is subjected to experience. Schopenhauer says something similar about the conservation of matter thinking this to be also a priori true. Boltzmann refers to some contemporary experiments that might contradict this theorem, and while he is convinced that these experiments are in error, he is even more convinced that this question must be decided by observation. How would Schopenhauer respond to the recently discovered phenomenon of radioactivity,

© Springer Nature Switzerland AG 2019
R. Schiller, *Between One Culture*, https://doi.org/10.1007/978-3-030-20538-6_21

he meditates. (Curious! Neither Schopenhauer nor Boltzmann realized that Lavoisier was the first to declare the theorem. Before 1789, was this not a priori knowledge? It was no knowledge at all!)

The extension of the notion of will to inanimate objects (*"when a stone falls to the ground this is as much an act of the will as when I myself will something"*) is the ultimate point for the physicist's patience. He is going to follow the philosopher even less in the final conclusion that life is nothing else but suffering. He keeps rather with Darwin and explains how the experiences that were found correct had been inherited in the course of the development of the human race and how the erroneous ones were discarded.

Boltzmann is not very amiable in the last sentence of this essay: *"Men would be freed from the spiritual migraine that is called metaphysics."* The audience of these lectures could have heard about a good many other subjects. Boltzmann's lecture notes have been preserved in Vienna, some of them being most detailed, other ones containing a few sentences only. One of them is just one word: *"probability"*—that is all that is written on the page.

Certainly, in that case he did not need any detailed exposition. This lecture must have been the summary of his most important thoughts, most far-reaching ones, which influenced the subsequent development of science as a whole. (Did the Viennese ladies know, visiting the lectures as a social event, that what they listened to was not the entertaining chat of a kind old professor but it was the Spirit by whom they were addressed?)

The problem was millennia old. Atoms? In the nineteenth century, the question was revived by chemists rather than by physicists. Chemists were mostly happy with their balls and sticks representing atoms and valence bonds, which form molecules if put together. Only few serious physicists supported these models. Atoms and molecules, if they exist at all, must be very small and incredibly numerous, even in a small portion of material. How to prove the existence and how to describe the motion of entities that many and that small? The direct proof of their existence came indeed somewhat later (four years too late, as we will see), but their motion, their velocities were successfully treated by something which first appeared to be only an ingenious trick of mathematics. Maxwell was the first to ask the correct question and he himself could give the correct answer. Let the individual fate of the atoms be forgotten and instead let the properties of their multitude be investigated. The appropriate question is: which fraction of the atoms of a gas move with a certain velocity? Maxwell found the answer, which even he thought was nothing more than a brilliant exercise in mechanics.

Boltzmann went a step further. A basic one. Heat, atoms, motion—his aim was to construct a general method for finding the laws they obey. Atoms always have different energies in a multitude, the actual energy of any given atom being a matter of *chance*. In order to obtain a law of the chance events, he introduced the notion of *thermodynamic probability*. This tells us how the energy is distributed among atoms, i.e. how many atoms have a certain energy (talking in very loose terms). The basic tenet by Boltzmann is that whatever thermal or chemical process takes place, *thermodynamic probability always increases and as the process is finished the probability attains its maximum*. This sounds very mathematical and indeed, according to Boltzmann, "*the problems in the mechanical theory of heat are at the same time problems in probability calculus*". Also it sounds almost obvious: as time goes by, conditions become more and more probable. But, obvious as it may appear, it has little connection with probability calculus, since the idea of probability varying with time is completely alien to mathematics. Boltzmann's tenet is a new law of physics: being not a theorem deduced from known axioms of mathematics but an independently recognized law of nature.

The value of any theory lies with its application to phenomena and observations. Everyone knows that if a cold and a hot body get in contact, the cold one gets hotter and the hot one colder; this never happens the other way round. Also, everyone knows that when dropping some wine into water, they mix by themselves and never demix spontaneously. These processes and many similar ones proceed in one direction only. As a matter of fact, the majority of natural processes are like that. Their directions are the same as that of the increase of their thermodynamic probabilities.

These were the ideas that were set forth in Boltzmann's popular lecture. The leitmotifs were *chance* and *probability*. Quite new ideas in physics, the laws of which referred to certainties in previous ages. The sun rises, stones fall—certainly they do. Now it turned out that such certain statements describe only a small fraction of natural phenomena. The majority of them obey Boltzmann's law. Einstein could prove the existence of atoms along Boltzmann's ideas, thus ending the atomic debate. This happened four years after the end of Boltzmann's life. Some twenty years later, although in a different context, probability became the basic notion of physics with the birth of quantum mechanics (Fig. 21.1).

Chance and probability instead of certainty took over their reign also in the arts and literature. Take for example *Picasso*'s painting. No definite action, no certain aim, the persons just linger around by chance. This is echoed in the first passage of Rilke's poem, the Fifth Duino elegy.

Fig. 21.1 *Picasso*: Strolling players with child, painted in 1905, a year before Boltzmann's death at Duino. This picture inspired *Rilke* to write the Fifth of the Duino elegies

> *But who are they, tell me, these Travellers, even more*
> *transient than we are ourselves, urgently, from their earliest days,*
> *wrung out for whom – to please whom,*
> *by a never-satisfied will? Yet it wrings them,*
> *bends them, twists them, and swings them,*
> *throws them, and catches them again: as if from oiled*
> *more slippery air, so they land*
> *on the threadbare carpet, worn by their continual*
> *leaping, this carpet*
> *lost in the universe.*

I wonder if Picasso or Rilke ever heard of Boltzmann. Nevertheless, both of them became good pupils of his.

Bibliography

Boltzmann L. On a Thesis of Schopenhauer. In: McGuinnes B, editor. Theoretical Physics and Philosophical Problems. Dordrecht: Reidel; 1974.

The poetry of Rainer Maria Rilke, selected poems, Duino Elegies and The Fountain of Joy. Kline AS, Translator. A commentary on the Elegies. https://www.poetryin-translation.com/PITBR/German/Rilke.php#anchor_Toc509812219

22

A Novel in the Phase Space

The reader's task seems to be easy with this novel by Max Frisch. The title, *Homo faber*, directly tells us what we are going to read; it is either about our distant ancestor, the Toolmaker, or—more probably—about a present-day descendant of the old fellow. So it is, the first-person narrator, an engineer is most proud of his knowledge in engineering in two ways: while moderately self-assured in technical matters, he is proud of his ignorance in any of the arts. He boasts loudly of being lowbrow and also rough when it comes to emotions. While reading the book, we might reasonably be surprised by such statements—rough and lowbrow, all right but, if so, why is he so self-aware and self-critical and how is he apparently most adroit in his vivid descriptions of nature. Anyway, let's listen to Herr Faber (this is his family name, indeed)!

Soon he tells us about his early and still existing interest in probability theory and statistical mechanics. Even a list of references is given by him: Reichenbach, Whitehead and Russel, von Mises … The subject of his thesis, a work never to be finished, would have been in this field discussing the problem of the Maxwell demon.

This demon was imagined by Maxwell in order to visualize the probabilistic content of the laws, which prevail in systems made up of very many particles. In short, our general experience is that if two bodies, for example gases in two separate tanks, are in thermal contact with each other and one of them is warmer than the other, the warmer one gets cooler whereas the colder one gets hotter. This is always so. But! A very small and very smart demon, if it existed, could alter this scenario. The particles in the warmer gas move faster than in the colder *on the average*. Nevertheless, slow and fast particles exist in both tanks; only their relative amounts are different. The demon can see the

© Springer Nature Switzerland AG 2019
R. Schiller, *Between One Culture*, https://doi.org/10.1007/978-3-030-20538-6_22

particles one by one and can select them by velocity. If there exists a small door between the tanks, the demon could decide to open it for a fast particle of the cold tank to make it fly into the warm one and also a slow particle of the warm tank to make it fly into the cold one. Hence, the cold tank becomes colder, the warm one warmer. However, having a huge amount of particles, as is the case with macroscopic bodies, let us say with a cubic millimetre of gas, a one by one separation is clearly impossible. Averages decide. Multitudes are governed by the laws of probability.

Human beings are also very numerous in the world; mankind is to be regarded as a multitude. Hence, Herr Faber could expect that … No! Anything that happens to him in the course of events is against chance, against probability calculus. His last weeks, which determine his life and predict his death, are controlled by the Maxwell demon. When flying from New York for one of his usual business trips, his plane makes an emergency landing in the Mexican desert. It turns out that the unpleasant passenger sitting next to him is the younger brother of a good friend of his whom he lost track more than twenty years before. During a chance chat, he also learns that this friend has married the love of Faber's youth. Later, this friend is found dead at a Guatemala plantation where the two men travelled rather at random. Leaving the younger brother in Guatemala, he returns to New York. Travelling to Europe by ship, he meets a young girl on board to leave her in Paris for good only to meet again a day later. They travel together to Italy and Greece and that is the place where the girl reveals her mother's name: she is the love of Faber's youth. Somewhere at the seaside near Corinth, the girl is stung by a snake; with difficulties he carries her to Athens where he meets the girl's mother and finally understands what he should have known before—he is the girl's father. The girl, despite excellent statistics on antidotes, dies. At the end of the novel, Faber is about to undergo an operation for stomach cancer. These operations, he knows, are successful with a probability of 96.4%, still…

Everything that happens is against expectations because—as the characters stress it repeatedly—probability calculus is unable to predict individual events. There is, however, something more than that the novel can tell the reader in connection with statistical thermodynamics. It is not clear whether this was stumbled upon by Max Frisch and Faber unwittingly.

Boltzmann's ideas on statistical thermodynamics were attacked by Ernst Zermelo, a contemporary mathematician, who had one serious argument. He could prove (successfully) that a system controlled by chance sooner or later returns to its initial state. That is, statistical laws do not account for processes which proceed only in a certain direction. The experience that warmer gas gets cooler and the cooler warmer cannot be understood in terms of statistics because at some later stage the opposite must happen: at first the differing temperatures become equal but later the initial temperature difference will be readjusted. According to Zermelo, Boltzmann's mathematics demands that in the space of events, in the phase space as it is called in the language of statistical physics, systems would go back to square one. This contradicts our immutable experience; hence a theory based on chance and probability must be in error. Boltzmann answered saying "sooner or later"—sure, but given the immense amount of molecules in a tiny, still macroscopic volume of gas, this takes more time than the age of the Universe.

This theorem is somehow part of the experiences of the protagonists in the novel and they must see that Boltzmann is right: there is no return to the origin in the phase space. At the end of the novel, Faber covers his previous route, for some reason or another, finding a good many things to be the same as before; the events are still not to be repeated. Returning to the beach of Corinth, he finds the girl's belongings; his travel companion, however, is not the girl but her mother. Flying back to New York, he is met by friends and colleagues but is unable to open the flat of his own and looks for his girlfriend in vain. He flies and drives to Guatemala just to find his friend's brother half-dead with heat and isolation. Finally, he returns to Athens to understand his fatal diagnosis. The geographical space can be recovered; the space of the events cannot.

This cruel truth is clear for the narrator at the end of the book. "*Even suicide could not alter the fact that once I was here in this world whereas at this moment my only desire is not to have ever been existed.*"

What has happened can never unhappen: a finding known of course for ages, even from theology; it is also an irrefutable result of statistical physics. *Zermelo's statement was falsified by Max Frisch.*

Bibliography

Frisch M. Homo Faber. Bulloc M, Translator. London: Abelard-Schuman; 1959.

Part VI

Fine Arts

23

Sculpting and Chemistry

Already in ancient Greece…

Hephaistos forged a shield for Achilles. The shield, as we all know, became a masterpiece, Homer's *Iliad* leaves no doubt about that.

> *He wrought the earth, the heavens, and the sea; the moon also at her*
> *full and the untiring sun, with all the signs that glorify the face*
> *of heaven-*

And he wrought also a good many scenes and views of the world doing it in such a marvellous way that the armour amazes the eyes of all who behold it. Being a real artist, Hephaistos knew where to start with his work.

> *Went to his bellows, turning them*
> *towards the fire and bidding them do their office. Twenty bellows*
> *blew upon the melting-pots, and they blew blasts of every kind, some*
> *fierce to help him when he had need of them, and others less strong*
> *as Vulcan willed it in the course of his work. He threw tough copper*
> *into the fire, and tin, with silver and gold; he set his great anvil*
> *on its block, and with one hand grasped his mighty hammer while he*
> *took the tongs in the other.*

He could set himself to his art only then. The immortal sculptor had to begin by alloying bronze, that is with metallurgy, chemical technology. His work was not to be imagined in any other way. In want of any appropriate metal, he could not even conceive his work, let alone to execute it. The artist

© Springer Nature Switzerland AG 2019
R. Schiller, *Between One Culture*, https://doi.org/10.1007/978-3-030-20538-6_23

is both master and slave of the material: he can frame it by his own will but knows only too well how far the material is willing to obey.

Art historians regard the appearance of bronze as an important turn in Greek sculpting. Until then smaller figures were moulded of clay, the larger, perhaps monumental, ones were carved of limestone or marble. Whereas these substances were easy to shape their strengths limited the forms of the figures. The sculpting of the archaic age (until around the sixth century B.C.) was in good accord with the properties of stone. The rigid and motionless, frontal figures stemming from the Orient had become milder by the chisel of the Greek master; nevertheless, the soft and grainy limestone prevented the sculptor from creating broad gestures, wide open arms. The two (beautiful) statues from Athens, reproduced here (Figs. 23.1 and 23.2), are very cautious with their motion and stand mostly peacefully. That was what the archaic sculptor could and would represent. This makes the result somewhat unexciting.

In the Acropolis Museum, Athens, about a dozen of kores flock around a statue of Pallas Athene[1] not all of them coming from the same workshop, all of them beautiful, all very similar.

Bronze alloy was known in archaic Greece as it is witnessed also by Homer. Tools and arms were forged out of it or figures were hammered as Hephaistos did. Sometimes wooden figures were covered with bronze plates. The great advantage of the alloy, however, had become patent with its casting. Bronze has a lower melting point than copper so it solidifies slower when cooled. Its viscosity being low it fills the mould completely. It is of high tensile strength, which enables one to make bulky figures without running any risk of deformation or break. Its properties and colour are easy to vary by changing the proportion of the components: whereas the most common alloy consists of 90% copper and 10% tin, the amount of the latter might vary between 4 and 32% sometimes with an addition of also zinc, lead or arsenic in order to change the colour. Initially, solid statues were cast; these, however, were heavy and demanded a huge amount of metal prompting the development of hollow casting or lost wax (*cire perdue*) process.

The process starts with the creation of the statue in clay which, after perfection, is covered with a thin layer of wax so carefully that the wax strictly conserves the form in clay. This is coated again with clay, the outer shape of the coating being immaterial. The internal and external clay figures are nailed to each other. Some openings are made on the external coating and the clay gets fired so the wax flows off through a lower opening leaving a hollow between

---

[1] This was so in the old museum, the cosy company having been disbanded in the splendid new one.

Fig. 23.1 Archaic Greek statue of a kore (girl) of limestone

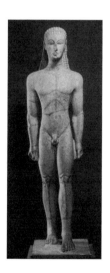

Fig. 23.2 Archaic Greek statue of a kouros (boy) of limestone

the bodies of clay. That hollow is filled with melted bronze. After its solidification, the external clay coating is broken off the cast, whereas the internal core is scooped out of it; the bronze statue is thus completed. Some refining and polishing may still appear necessary; nevertheless, the artist's real work is done in clay. Larger statues are cast in parts which are put together afterwards; in such cases, special care must be taken of the composition of the alloy in order to ensure a uniform colour of the separately produced parts. Sometimes pieces

Fig. 23.3 Bronze statue of Zeus or Poseidon

of lead are also put inside the statue in order to make it stand safely. The positive core gets annihilated by this method later, however, it was refined so as to preserve the core, making possible the multiplication of the statue.

The huge statue representing Zeus or Poseidon, shown here (Fig. 23.3), was obviously produced by hollow casting. It was recovered at cape Artemision in the muddy seabed among the load of a submerged ancient ship in the twentieth century. (One of its arms, cast independently, was found only later.) The bold movement, the god's left arm stretched out, the right one throwing a spear or harpoon could be realized only in bronze. It is a lucky exception to have this statue extant. Only few of the bronze figures survived since the alloy was of great value, and when there was a need (quite often there was some), the figure was melted to produce arms. That is the reason why most of the Greek sculptures are known only in Roman marble copies. Poseidon's heroic gesture, however, would have been impossible to translate in marble.

History of a Failure

French ceramicist *Bernard Palissy* lived in the sixteenth century, between 1510 and 1590. He is said to have elevated rustic pottery to the height of Renaissance art. This happened against his own intention.

A learned, much travelled person, ceramicist, portrait-painter, glass painter and surveyor, he was already married and settled when he got hold of a white cup, finely enamelled. Was it a Chinese or an Italian make is not known.

Fig. 23.4 Bernard Palissy: Salt tray

Palissy made up his mind to ferret out the secret of its production. Despite all his experience, he felt "*to fumble in the dark*". Having spent sixteen years performing strange and even stranger experiments, he impoverished together with his family, finally being reduced to split his own furniture and flooring in order to heat his furnace, still was unable to produce the supremely subtle substance. In the end, the French duke *Montmorency* saved him in his utter misery and despair seeing, of course, not the substance that never came into being, but his simple, harshly coloured, sturdy vessels shaping animals and plants in a naturalistic manner (Fig. 23.4). These works of his made him rich, famous even at the Royal Court and, a faithful Protestant, protected him from religious persecution during the savage times of the Saint Bartholomew's night massacre. (At least for some time. At the age of eighty, he was imprisoned, and although the King was ready to pardon him in return for catholicizing, he refused apostasy and died in the Bastille, thus avoiding execution.)

Palissy was a notable person not only as an experimental chemist but also as author, lecturer in popular science and above all as a rational thinker. His argumentation against alchemy is subtle and convincing. Some historians think his literary production is more important than his rustic pottery; nevertheless, the latter ones were regarded as the apex of naturalism in the Renaissance by the great Dutch art historian Huizinga.

Palissy failed to find the secret of the wondrous substance. He used well-known substances: clay, everyday enamels and colouring agents; probably he even did not use a potter's wheel, these simple materials and methods leading or rather compelling him to a crude, in some cases fantastic, naturalism. At the beginning, he fought the substance and lost. Later, he obeyed the substance.

A Portrait

Marcellin Berthelot was one of the greatest experts of nineteenth-century organic and physical chemistry of France. It is certainly not the task of the present writing to give an overview of his activities this place being much too limited to summarize his achievements, even in the form of a table of contents. He was incredibly prolific both at the bench and the desk. Being a leading authority also beyond the spheres of science, he did his best to cultivate this authority. Legend maintains that he cut a small opening in his laboratory apron, a usual wear of his period, in order to make visible, all the time and for everybody, the rosette of the Légion d'honneur he wore on his lapel. Certainly, it was one of the dense rain of acknowledgements and distinctions that Rodin created a sculpture of his (Figs. 23.5 and 23.6). It would be interesting to know what these two great men knew and how they thought of each other. Most probably, Berthelot held the sculptor in esteem, having dedicated several

Fig. 23.5 Photograph of M. Berthelot

Fig. 23.6 August Rodin. Portrait of M. Berthelot

of his writings to him. However, what sort of information did Rodin gather about his work? And what do the present-day visitors of the Rodin Museum in Paris know?

Bibliography

Homer, The Iliad. transl. Butler S. http://classics.mit.edu/Homer/iliad.mb.txt
http://headforart.com/
https://www.strangescience.net/palissy.htm

24

A Reader's Diary About Humanist Friendship

Drawing a world map is quite a troublesome task. One ought to unfold the sphere on a plane and this is, as we all know, impossible. That is an age-old problem and so are the approximations cartographers have applied for its solution. A number of projections have been developed in order to transform the sphere into a surface which can be unfolded. All these approximations go with inevitable loss of information and the obvious question refers to the type and extent of distortion which are still compatible with the cartographer's practical aim. It was the era of the great discoveries when the interest in this problem was renewed.

Perhaps *Mercator*'s method is the easiest to visualize: here the surface points of the sphere are projected to the side of a cylinder which surrounds the sphere. Having folded out the cylinder, the areas appear immensely large near the poles in comparison with the neighbourhood of the Equator and the poles cannot be represented at all; nevertheless, the map's main advantage is that it conserves the angles i.e. the directions on the map are the same as in reality.

A different approach is needed if the relative areas of certain pieces of land are to be recorded. For that purpose, the sinusoid or *Sanson-Flamsteed* method has been applied. The latitudes are represented by parallel straight lines whose lengths are proportional to the latitudes, whereas the meridians appear as curves which meet at the poles, intersect the Equator in right angle and their lengths increase with increasing latitudes (Fig. 24.1).

Another treatment developed by *Werner* is related to the above method. The latitudes obey the same rule, but the whole transformation is more involved in order to achieve the conservation of both areas and distances. Neither the Equator nor the meridian in the middle of the map is distorted.

© Springer Nature Switzerland AG 2019
R. Schiller, *Between One Culture*, https://doi.org/10.1007/978-3-030-20538-6_24

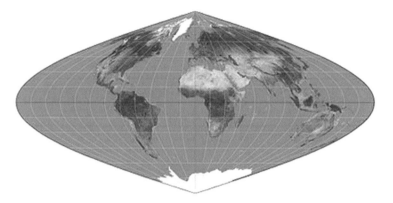

Fig. 24.1 The sinusoidal (Sanson-Flamsteed) projection

Fig. 24.2 The cordiform world map by Honterus

The result is a cordiform (heart-shaped) figure, one of its oldest examples (and a very fine one) having been drawn by Johannes Honterus in Transylvania, 1546 (Fig. 24.2).

For some time, this heart shape was a favourite with the cartographers. *Abraham Ortelius,* the great humanist, art collector, publisher, merchant and cartographer, published a world map of similar shape in 1564. Practice seems to have preferred other solutions because from 1570 onwards he abandoned the heart shape. The new map (Fig. 24.3) was a great success, being republished

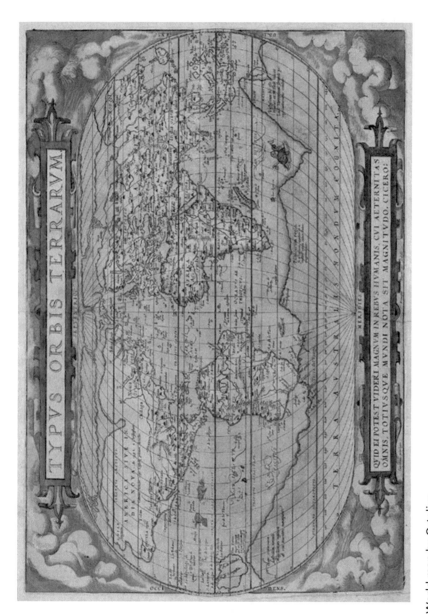

Fig. 24.3 World map by Ortelius

several times over the next decades. This work is also a proof of the sincere friendship between two scientists: Mercator was of great help to Ortelius with determining a number of positions.

Friendship among humanists, the republic of letters, was of utmost importance in these times and in the Low Countries perhaps even more so than anywhere else in the world. Dutch revolt, religious disturbances, Duke Alba's campaign—a good number of reasons which have induced the scholarly spirit to become secluded in its own realm in order to find protection at least by friendly words at a time when the surrounding world argued with lancet, arson and the stake. A written document of this disposition is still extant.

Similar to other humanists, Ortelius started to make a collection of citations, dedications and signatures in 1574. Finally, this *album amicorum* (album of friends) contained 130 entries. Mercator, *Lipsius* the philologist, *Jean Bodin* the lawyer … men of different ages and social standing (even a woman, which was no small wonder), Catholics and Protestants—their humanist lifestyle, their predilection for the written word, the much-cited Latin *auctor*s wisely and cosily kept them apart from the outer world which was both dangerous and to be detested. *Contemno et orno*—I despise and praise (this world); that was Ortelius' motto.

The desire for common meditation brought about spiritual communities in the shadow of the great religious movements and warring churches in an age when personal fate was greatly influenced by one's religious affiliation. Such communities usually worked in secrecy since an open appearance would have been deadly dangerous. One of them was the clandestine movement called *Familia Caritatis,* the Family of Love, which—as far as one knows—Ortelius was also a member of.

The Family (I refrain from calling it either a religion or a sect) was founded by *Hendrik Niclaes* around 1540. His mysticism rested upon the belief that every human being inherits a part of Divinity. The idea, with all its corollaries, became mostly popular with artists, scientists and educated people as a whole. Probably they did not regard themselves as heretics since they all lived in peace in their Catholic or Protestant faiths. Wisely they avoided any clash with ecclesiastical or secular forces. The members of the Family, guided by some hidden spirituality, tried to find a direct connection with God. By this, however, they recovered the old heretic proposition on man's double nature—human and divine. Their tolerance was extended also to Jews and Muslims because they maintained that everyone can follow the same path of esoteric elevation. The symbol of this kind of spirituality was the human heart. Did geometry and religion meet on Ortelius' desk?

An important member of the circle of friends and probably also of the Family was the great painter, Pieter Bruegel the Elder. Called often Peasant Bruegel due to his numerous pictures of folk scenes, his landscapes are most telling about his relationship to Nature. As the art historian Charles de Tolnay writes "Nature as a single organism is permeated by the 'Weltgeist' (spirit of world) thus intelligence is revealed in each and any being." Bruegel's landscapes often appear as living creatures.

Martin Kemp published a paper in *Nature* about the friendship of Ortelius and Bruegel some years ago. The two great men, he writes, tried to complete Ptolemy's two-sided program of the geographic and chorographic description of the world. Whereas geography considers the world as a whole, chorography deals with smaller neighbourhoods, like cities and ports. The author of the *Theatrum Orbis Terrarum* was interested also in the history, economy and society of the described larger areas; the painter's attention, however, was fixed to the moment and site of a scene.

The wooden plough, the ship leaving the port, the shepherd gazing into the thin air do not merely decorate his well-known picture, *Icarus* (Fig. 24.4). They *are* the picture. The furrows in the ground and the folds of the ploughman's robe are oriented as the meridians and latitudes on a map. Now, forgetting all reverence towards an immortal art work, we can make a mirror image of the picture through its left side and put the original picture and its mirror

Fig. 24.4   P. Bruegel the Elder: *Icarus*

Fig. 24.5 Icarus with its mirror image

Fig. 24.6 P. Bruegel the Elder: *Icarus, detail*

image next to each other; that way we obtain Fig. 24.5. The figure has a definite resemblance to a cordiform map, for example to Honterus' work. And by the same token, it is also the symbol of Familia Caritatis.

The titular subject of the picture appears only as a pair of legs floundering in the water—if one recognizes it at all. Even if this small detail is enlarged (Fig. 24.6), the fisher, the waves and the branch over the water look more important than the drowning man. At first sight, the painter seems to follow

Ovid's poem, but no! According to the poet the fisher, the shepherd and the ploughman are astonished by the event. Not in the picture, here they pay no heed to what is happening.

Was that the wise and sober way the Family of Love looked upon worldly matters?

Bibliography

Harris J. The practice of community: humanist friendship during the Dutch revolt. Tex Stud Lit Lang. 2005;47(4):299–325. http://muse.jhu.edu/journals/texas_studies_in_literature_and_language/v047/47.4harris.html

Kemp M. Looking at the face of Earth. Nature. 2008;456:8760.

Mangani M. Ortelius's Typus Orbis Terrarum (1570). 1952. http://www.giorgiomangani.it

de Tolnay C. Die Zeichnungen Pieter Bruegels. Zürich: Rascher Verlag; 1952.

25

Victorians Looking into the Skies

A visit to England's Lake District—an experience to be repeated and a pleasure without an end for the literati of Great Britain and Continental Europe. At least for those who still adore early romantic Lake Poetry with its inclination towards the countryside and worshipping of nature. Quite numerous they seem to be even nowadays, wandering among the vales and hills around the lakes.

My visits in that area had a very different motive. Chemists active in a field of research called radiation chemistry have their regular gatherings in that area. As one of their number, it happened several times that I could utter expressions like "electron solvation", "geminate recombination" or "double strand break" in a lecture room on a hilltop at Lake Windermere. I am not going to divulge the meaning of those arcane expressions, mainly because I presume there are but few who would not find the whole subject rather boring. There is, however, something else that makes the Miller Conferences (that being the name of the gatherings) enjoyable. With lectures held in the mornings and early evenings, the afternoons were left free for informal discussions or excursions in the neighbourhood.

On one occasion, we went to Rydal Water to visit Wordsworth's Dove Cottage. A definite break from our daily chores. This sacred shrine has indeed had little to do with our cherished morning tasks. "*Enough of Science and of Art*" Wordsworth cried in one of his poems. And in a prose passage he wrote: "*The knowledge both of the Poet and the Man of science is pleasure; […] The Man of science seeks truth as a remote and unknown benefactor […] Poetry is the breath and finer spirit of all knowledge;*" The footsteps of Coleridge, a more scientifically minded visitor, are obviously not exhibited. The only item which could

© Springer Nature Switzerland AG 2019
R. Schiller, *Between One Culture*, https://doi.org/10.1007/978-3-030-20538-6_25

Fig. 25.1 Tomb of John Ruskin at Coniston

recall our profession was the chemist's scales used by De Quincey for measuring the correct helping of his daily opium.

Another day, walking at Coniston Water, in the churchyard, we came across the gravestone of the great mind of a later generation who taught his contemporaries the irrevocable laws of how to enjoy a picture or a sculpture, or more generally: what is the use and the abuse of art. We were standing at John Ruskin's grave. The Anglo-Saxon type cross is richly carved with symbols referring to Ruskin's literary work. However, a mere glance at the cloud-covered sky brought back a good part of his age and his way of thinking (Fig. 25.1).

Ruskin was a contemporary and a fervent advocate of the Pre-Raphaelite movement, which tried to rejuvenate English art by denouncing everything that remained Baroque and returning to early Renaissance and Gothic ideals. All these happily coincided with his religious beliefs, his moral stance and social conscience. His often disclosed love to nature, however, must have been in need of something closer to real experience than the movement's pale symbolism. Turner's landscapes and seascapes had a great appeal to him. Fog lying

Fig. 25.2   J.M.W. Turner, Quillebeuf, at the Mouth of Seine

on the fields or above the sea, dust, steam and smoke on a railway line, ragged, thick clouds in the skies, these were the subjects in Turner's pictures which apparently attracted him the most. A forerunner of Impressionism? Might well be, but also someone who could express the cloudy disposition of a most successful age. Spleen, together with the sombre sight of smoking factory chimneys, the latter also making Britain prosperous and mighty, however (Fig. 25.2).

In the second part of the nineteenth century, Englishmen were interested in the clouds, literary gentlemen and scientists alike. Ruskin wrote a very detailed essay about "*The storm-cloud of the nineteenth century*" giving the natural history of clouds by cataloguing their origins, appearances and effects on human emotions. Praising the by-gone beauties of the skies of his youth, he is busy to blame the "*plague-clouds*" (an early synonym for smog?) being belched by industrial plants.

In these years not only Turner, the painter and Ruskin, the art critic were investigating the clouds but also John Tyndall, a mountaineer, scientist and populariser of science. A professor at the Physics Department of the Royal Institution, a venerable establishment for the furthering and dissemination of science since the days of Davy and Faraday, he became a well-known and highly appreciated figure among the men of science in Britain. An excellent observer, he was the first to describe a phenomenon originally called Tyndall

Fig. 25.3 Tyndall effect

effect, later carrying the name of Rayleigh who gave its rigorous theory. Tyndall realized that light is scattered when it traverses a gas or liquid in which very small droplets or particles are afloat. Scattering means, in the physicists' language, that whereas most part of a light beam propagates forward, a smaller fraction of it can be observed sideways as it is shown in Fig. 25.3.

Not a theoretician, Tyndall offered a qualitative explanation only, which, however, turned out to be sound and valid. It was based on the wave theory of light and analogies with waves on liquid surfaces enabled him to understand what he saw. His way of description is pictorial, making his ideas easy to follow. "*A small pebble, placed in the way of the ring-ripples produced by heavy rain-drops on a tranquil pond, will throw back a large fraction of each ripple incident upon it, while the fractional part of a larger wave thrown back by the same pebble might be infinitesimal. Now to preserve the solar light white, its constituent proportions must not be altered; but in the scattering of the light by these very small particles we see that the proportions are altered. The smaller waves are in excess, and, as a consequence, in the scattered light blue will be the predominant colour.*" Light is scattered by the tiny floating particles as water ripples are scattered around a stone, and the wavelengths of the scattered ripples and of scattered light are changed in the same sense. He realized that a most important phenomenon in nature can be explained by this finding. "*I am unwilling to quit these chromatic phenomena without referring to a source of colour which has often come before me of late in the blue of your skies at noon, and the deep crimson of your horizon after the set of sun.*" That was the idea which enabled Tyndall to explain why the sky is blue and the setting and rising sun is red.

Ruskin knew Tyndall's work and was furious with it. "'*When,' says Professor Tyndall, 'you are told that the atoms of the sun vibrate at different rates, and*

Fig. 25.4   Turner: The Shipwreck—detail

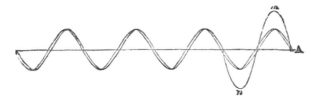

Fig. 25.5   Tyndall: Light wave

produce waves of different sizes,—your experience of water-waves will enable you to form a tolerably clear notion of what is meant.' 'Tolerably clear'—your tolera-tion must be considerable, then. Do you suppose a water-wave is like a harp-string?" Ruskin does not seem to have ever understood the relation between vibration and wave motion. Anger by ignorance. His picture of a wave must have been that of a romantic painter (Fig. 25.4).

A physicist, like Tyndall, has had a less poetic but a more sober and math-ematically well-defined idea about a light wave (Fig. 25.5).

Ruskin harboured somewhat complacent ideas about science: *"I do not mean by science, for instance, the knowledge that triangles with equal bases and between parallels are equal, but the knowledge that the stars in Cassiopeia are in a form of W."* He preferred description to understanding. It is admittedly childish to scold a great mind who departed more than a century ago but I cannot suppress a late piece of advice—perhaps learn some science? But Ruskin knew the importance of learning whenever he was within his own realm. One cannot expect a weaver from Liverpool or a miner from Manchester to adore a picture by Botticelli in his leisure time—he wrote. Sure, because some education is needed for enjoying art. What about enjoy-ing science?

Tyndall was more permissive with art than Ruskin was with science. He concluded an address given before the British Association with these ideas:

"The world embraces not only a Newton, but a Shakespeare—not only a Boyle, but a Raphael—not only a Kant, but a Beethoven—not only a Darwin, but a Carlyle. Not in each of these, but in all, is human nature whole. They are not opposed, but supplementary—not mutually exclusive, but reconcilable."

I do hope Professor Tyndall is right.

Bibliography

John Ruskin. The storm-cloud of the nineteenth century. www.wwnorton.com/college/english/nael/noa/pdf/27636_Vict_U08_Ruskin.pdf

John Ruskin. The eagle's nest. http://www.gutenberg.org/ebooks/42917

John Tyndall. Six lectures on light. www.gutenberg.org/files/14000/14000-h/14000-h.htm

John Tyndall. Address delivered before the British Association assembled at Belfast: with additions. https://archive.org/details/addressdelivered00tyndrich/page/n6

www.science20.com/florilegium/blog/wordsworth_science_and_poetry

26

Point-by-Point

"*We paint as the bird sings*"—said *Monet* enunciating what could be called the impressionist painters' practical credo. Standing in front of their pictures we believe him. The Hungarian impressionist *Rippl-Rónai* supported this idea of "*painting at once*" with a historical argument referring to the frescoes of Masaccio or Fra Angelico. These are seen to be fresher than the panel paintings of the same masters since the artist is compelled by the technique of fresco painting to paint at once, obviously not the entire work but detail after detail.

In contrast to the relative freedom of preliminary considerations and technical manipulations praised by Monet, a contemporary group in Paris, calling themselves neo-impressionists, introduced a slow and cumbersome technique that came to be known as pointillism. Instead of using bold brush strokes, their pictures were constructed from extremely minute spots which finally revealed shape, colour and light, the process of painting being a matter of several months of thoroughly devised hard work. Curiously enough, the movement was not an offspring of some spontaneous, internal development of art but was born of the study of scientific literature.

Michel Eugène Chevreul was one of the great chemists of the nineteenth century. He lived a long life. As a child, he still could have seen *Lavoisier* and died about the time when *Marie Sklodowska* arrived in Paris. Among other important results of his, he discovered the structure of fats, separated a number of fatty acids, isolated cholesterol from gallstone and was active in the soap and candle industry; in short, he enriched our knowledge in organic chemistry with important results and observations.

© Springer Nature Switzerland AG 2019
R. Schiller, *Between One Culture*, https://doi.org/10.1007/978-3-030-20538-6_26

At a stage of this long and active career, he was in charge of the paint and colouring production at the Gobelin manufacture in Paris. Complaints about the qualities of the materials were frequent the weavers finding the colours of the fibres to be much too faint. According to Chevreul's investigations, the chemistry was all right but the near lying fibres of different colours influenced the resultant effect. What he found was summarized in a book entitled *The Law of the Synchronous Contrast of Colours*. Its first edition having been dated 1839, it was edited again fifty years later, at the height of pointillism.

Chevreul's idea had its forerunners. Already *Leonardo da Vinci* asked: "*How to compose colours so as to make them mutually more beautiful?*" Goethe's *Theory of Colours* must have directly influenced Chevreul. Having been unable to lay my hand on his original work, I had to do with extracts. Three components of the solar spectra are considered as basic: blue, red and yellow. Any pairs made out of them produce three further colours called doublets. The third colour not being the member of the doublet is its complementary. Complementary colours strengthen each other if they are neighbours but extinguish each other if their grains are mixed. Two doublets having one common component fade each other. The main statement is: "*Two colours being next to each other are seen to be as differing as possible.*"

The author had higher ambitions than the establishment of these simple rules. Having constructed a circle of 72 sectors, he tried to develop a general system of colour harmony, based on analogies and contrasts, which might be similar to the harmonies in music. These ideas did have some scientific consequences in the colour vision theory of *Helmholtz* and in *Ostwald*'s colour system, results of which are still in use. Moreover, some artists, like *Johannes Itten*, one and a half centuries after Chevreul, were willing to find similarities between optical and acoustic harmonies.

Neo-impressionists were serious and diligent in studying Chevreul's work and also that of his follower, Ogden Rood. (Who, once dropping into a pointillist exhibition, wished he had never written his book.) They were strict in obeying the rule that paints must not be mixed either on the palette or on the canvas; only spectrally pure colours are to be put next to each other and it is the spectator's retina or rather his/her mind where the sense of colours finally appears. The effect must be produced by optical and psychological means and not by the substances on the canvas. The idea of spectral purity was held so much in awe that brown and black were not in use at all, those being absent from the solar spectrum.

Seurat has been held to be one of the greatest artists of the group. His perhaps best known and most highly esteemed work, *A Sunday Afternoon on the Ile de la Grande Jatte,* was the result of two years of intense work (Fig. 26.1).

Fig. 26.1   Seurat: *A Sunday Afternoon on the Ile de la Grande Jatte*

Fig. 26.2   Edison with his movie camera

This has nothing to do with Monet's ideal, and the splendid result is obviously to the credit of the painter's talent and not of his diligence or theory.

Still, his theoretical stance deserves some attention: "*The light effects on the retina of the eye are temporarily defined phenomena.* [...] *The tool of expression is the optical mixing of shades and colours* [...] *according to the laws of contrast.*" Mixing and contrast—this is Chevreul's way of thinking. Clearly there is a difference between the physics of light, the chemistry of the dyestuffs and the psychology of colour sensing. The third one is the painter's field. There is, however, a remark also on time in the above sentence. Once it was thought that painting or sculpting are arts of space, whereas music is an art of time. Seurat is apparently at variance with this classification. Time does play a role in visual art, because seeing is a temporal process. Time is needed for the eye to wander from one spot to another; the period of impression is not infinitely short; hence the artist's creation can be grasped only slowly. Point-by-point.

That is our daily experience at the cinema or in front of the TV screen. Seurat died in 1891 at the early age of 32. It was the year when the movie camera was patented by Edison (Fig. 26.2).

Bibliography

Itten J. The art of color. New York: Wiley; 1973.

Sérullaz M. Phaidon encyclopedia of impressionism. Oxford: Phaidon; 1978.

Seurat (Einleitung von Jedding H), Emil Vollmer Verlag, Wiesbaden.

http://www.colorsystem.com/projekte/engl/17chee.htm

http://dbeveridge.web.wesleyan.edu/wescourses/2001f/chem160/01/Course%20
  Materials/Chapter%20Notes/chapter 5.doc

27

Icons by a Chemist

The Russian theologian, philosopher and art historian, Pavel Florensky definitely disliked renaissance painting. He scolded these painters: "*In order to show naturalistic pictures, a world which has got loose of God and Church, a world the laws of which rest in itself, they need strong and full-blooded sensuality. These pictures are sensory entities which have to prove their own existence the more emphatically the better.*" The orthodox Christian thinker rejected and despised everything what Western common opinion maintains about fine arts, notions which are thought to have been realized and perfected to the full in the age of Renaissance. The Russian icon is something completely different. It is not a creation of aesthetic value, it is not the production of the artist's imagination and talent. "*The iconostas is a vision.*" A well-defined vision. "*The iconostas is the congregation of the saints.*" Not the depiction of the congregation but the congregation itself. The icon is still a symbol, but if it is a genuine creation, "*it becomes united with what it symbolizes*".

Few people are blessed with the ability to see the gathering of prophets, saints and angels. It is rare to find even one of them in the populous guild of icon painters. The Russian Church thinks this is not necessary. "*The painter deals with technical matters only, the arrangement is decided by the Holy Fathers. Icons have never been based on independent ideas, their painters do not conceive their work but create them according to the scatheless laws and traditions of the Universal Church.*" As Rublyov painted his celebrated Trinity icon, his hand was controlled by saintly Nikon.

Talking here about science that much is enough for us about icon painting. Still, this introduction might prove to be helpful to make clear the nature of some interesting and beautiful objects. Dr. Béla Vizi, reader in chemistry at

© Springer Nature Switzerland AG 2019
R. Schiller, *Between One Culture*, https://doi.org/10.1007/978-3-030-20538-6_27

the University of Veszprém, Hungary deals, on the one hand, with the spectroscopy of molecules; on the other hand, he is a sculptor. This statement is somewhat misleading—not on one hand and on the other but in complete simultaneity. Vizi is a sculptor trained in science with a need to understand and represent both fields in full concert and mutual support. His exact knowledge in molecular physics and his flair in sculpting appear synchronously. In this vein, he wrote his scientific papers, gave his university lectures and created his figures.

The introduction of one of his booklets carries the title *Chemistry and fine art.* This is a thoroughly composed introductory text for the general public about the electron structures of atoms and molecules, the periodic table, structural formulae, molecular vibrations, the elements of reaction dynamics—a good introduction to quantum chemistry for the layperson. Fine art remains unmentioned.

The forthcoming section of the book reveals that the statues represent similar ideas. It is perhaps not much surprising. The chemist's way of thinking makes regular use of models, line drawings, graphic representations. One knows that in reality most of the pictures represent steric structures projected onto the plane.

In reality—what does this mean? There are a number of experiences in chemistry. Some substances burn, others evaporate, some of them are yellow or red, and there is a substance 16 grams of which reacts with 2 grams of another one—and innumerable similar observations. The chemist's task is to find out why it burns, why it is yellow and why exactly 2 grams. The explanations are usually of a mathematical nature. First, only elementary algebra was needed to express the weight relationships or the gas laws. Later, the differences between some compounds were explained in terms of complicated stereo geometry. The relationship between atomic weights and the properties of elements was understood in terms of an unexpected arithmetical rule. Molecular motion was described through classical mechanics and probability calculus, whereas their structures can be revealed only through quantum mechanics. Now, in order to make all these mathematics easier to digest, the results were tried to be summarized in drawings and mechanical models. Structural formulae were drawn, atoms were represented by balls sometimes connected by springs or forming clusters which are similar to bunches of grapes, complicated bodies of revolution were devised in order to visualize electron distributions.

No one has ever seen such objects in nature. They are mere tools for learning and understanding, since facts and ideas hold fast in our minds through visible and tangible objects. Vizi's sculptures, or at least a good part of them,

Fig. 27.1 Scent of a rose (by courtesy of the artist)

are the witnesses to learning and understanding in that way. The images do not stem from sensory experiences but express concepts. They become objects of art due to the artist's talent and his trained scientific mind. As a result, we are presented with attractive shapes made of alloys of noble lustre, of thoroughly stained wood or of well-chosen pieces of stone. There are some lucky cases where the primary chemical content is associated with some deeper meaning.

A good example is the small bronze entitled *Scent of a rose* (Fig. 27.1). The scent comes in part from the organic compound called phenyl ethyl alcohol, the structural formula of which is given below. The statue mimics the formula in an artistic way and at the same time it recalls the shape of a flower.

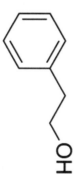

Phenyl ethyl alcohol

Fig. 27.2 The periodic table (Helios-Gaia) (by courtesy of the artist)

Fig. 27.3 The p⁴ electron configuration (by courtesy of the artist)

Another bronze entitled *The periodic table (Helios-Gaia)* (Fig. 27.2) carries a multifaceted meaning: chemical elements, as represented by small stars, line up in spirals (a reference to the Periodic Table); the spirals recall also the Cosmos by their similarity to the arms of a galaxy, whereas the three lobes may refer to the electron shells.

A series of statues represent electron distributions which are the results of quantum mechanical calculations (Fig. 27.3). Quantum mechanics is known to have little connection with everyday visual experience. What is called the shape of an electron cloud is a mathematical idea; it is the probability distribution of finding the electron in the neighbourhood of a given place. The solutions of partial differential equations are here visualized. Although static, the statues suggest even the dynamics of electron motion, something which can only be imagined.

Béla Vizi's hand was controlled by saintly Dmitry Mendeleev and Erwin Schrödinger.

Bibliography

Pavel Florensky P. Iconostasis, transl. Andreeva O, Sheehan D. St. Vladimir's Seminary Press; 1972, 1996.

Vizi B. Chemistry in sculpture. The Chemical Intelligencer. 1995, January.

28

Line: Exhibition Opening Speech

Having walked over the exhibition rooms, I had to see that the pictures are mostly about chemistry. Particularly about its most fundamental question, which determines how a physicist, chemist or astronomer thinks about nature.

The prime problem of science has been (and will be) the construction of the world in us and around us. Some simple, easy to understand answer has been looked for, since it would be most disappointing to believe that our world, complex and multifaceted as it is, lacks any idea of organization. There must exist some abstract law which can be extracted from the disturbing multitude of events and phenomena. Antiquity had two answers to offer, given almost simultaneously. *Empedocles* and the peripatetic school thought that the world consists of four substances: earth, water, air and fire. Everything that can be found, seen, smelt or felt in the sublunary world is a mixture of these four elements which differ only in their proportions in each material. Other thinkers, *Leucippus* and *Democritus,* maintained that every substance consists of indivisible and incorruptible particles, atoms, which are bound to each other with small arms or tendrils, and this is the way how different materials come into being. Looking back after more than two thousand years, it is clear for us that two answers were given to two questions, one of them referring to the composition and the other to the structure of substances. It took two millennia before the great scientist Robert Boyle could realize that the dichotomy lies not in the answers but in the questions.

Anyone who has ever enjoyed the dancing grains of dust in the sun rays or has let trickle a handful of sand through the fingers would find the existence of atoms quite natural. Indeed, it were not the naïve observers of nature but serious, critical thinkers who refused the idea. According to their reasoning, if

© Springer Nature Switzerland AG 2019
R. Schiller, *Between One Culture*, https://doi.org/10.1007/978-3-030-20538-6_28

atoms exist something that is non-atom must also exist. There must be voids between the atoms; otherwise they would be unable to move and the whole world would be static, at constant rest. But void cannot exist because it cannot be imagined. No, since void is thought to be nothing, however, "if someone says *void does exist* then void is *something* so it cannot be *nothing* at the same time,"—that was about the way how *Parmenides* refuted the idea of the void (of vacuum), consequently the idea of atoms and their motion.

It is perhaps no exaggeration to state that atomic theory, hence almost everything upon which our present-day scientific understanding rests, depends on the existence of voids, of empty spaces. An art gallery is perhaps not the appropriate place to discuss the development of the physicist's view on the difference between geometrical and physical vacuum, although this exhibition takes the viewer to the pure, unadulterated idea of the empty space. Because it is about the line.

A wood panel or a canvas is a well-defined part of the plane. If the artist paints a deer at a brook, the Queen of Cyprus or a non-figurative form, he or she separates a smaller part of the plane from the whole. The plain gets divided into parts which might be most artistic but the idea of a plane still remains as it was. Finally, the parts cover the whole plane encircled by the frame.

However, the situation is completely different if the painter draws only a line. The ideal, imaginary line has only length without having any width. It is a one-dimensional form on the two-dimensional plane. However, many lines (but still less than infinitely many) are drawn, the plane remains as uncovered as if not a single line were drawn. The line might be straight or curved, might

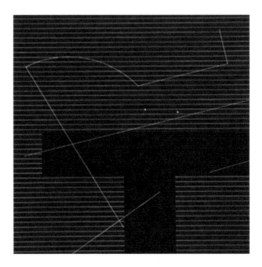

Fig. 28.1 Tamás Konok: Motion of lines (by courtesy of the artist)

bend, wind or straggle it is still a line. Whatever the shapes of two curves are there gapes a void of plane between them on the canvas. A two-dimensional void between the one-dimensional entities (Fig. 28.1). The non-line between the lines. An empty area the existence of which was denied. The lines which quiver, bolt, bow, bend on every picture of the collection circumscribe parts of the plane which are non-lines.

It is of course an abstraction that we are talking about. The track of the pencil or paint brush is not thin beyond imagination; in effect the line drawn by the artist has its width and is no one-dimensional part of the plane. But the artist's intention and the resulting piece of art compel us, viewers, to imagine the object as a line, an element which is unable to cover the plane. And this helps us to imagine the otherwise unimaginable, the existent *Nothing* between the existent *Somethings*. The void between the atoms.

Part VII

The Pleasure of Doubt

29

A Sentence Between Plato and Planck (On the 450th Anniversary of Galileo's Birth)

I am not going to write about *that particular* sentence. It is not known, anyway, if he uttered it indeed. A legendary word by which posterity wanted to remedy something that was thought to had been the deficiency of the genius' character. Yet it moves? *Galileo*'s spirit shall not be invoked by this doubtful anecdote!

The sentence to be quoted is in Galileo's *Discorsi*. The form of this last work of his is a dialogue similar to the earlier *Dialogo* where the Ptolemaic and Copernican systems were discussed and Copernicus' views were tried to be proved. As it is well known, it was this book that resulted in the trial, sentence and the forced abrogation of the heliocentric system. The characters are the same in both books: *Salviati* expounding Galileo's ideas and arguments; *Sagredo* doubting all the time, even though his way of thinking is akin to Galileo's; and *Simplicio* representing Peripatetic philosophy and down-to-earth common sense. The *Discorsi* consists of four chapters, i.e. "four days of the discussion". The first two days are devoted to the breaking and strength of solid bodies based on general ideas of the structure of materials, whereas in the last two days, the theory of motion is dealt with. This overview is only roughly true. Due to the dialogical situation, it is inevitable that basic problems like free fall or the motion of the pendulum should be discussed during the first day. Galileo, an old man now, summarizes his total physical world view in this work (obviously excepting those which refer to the solar system).

Simplicio maintains Aristotle's views on free fall, something which have not been doubted over two millennia: "*[Aristotle] supposes bodies of different weight to move [...] with different speeds which stand to one another in the same ratio as*

© Springer Nature Switzerland AG 2019
R. Schiller, *Between One Culture*, https://doi.org/10.1007/978-3-030-20538-6_29

the weights; so that for example, a body which is ten times as heavy as another will move ten times as rapidly as the other."

Salviati, however, *"greatly doubt[s] that Aristotle ever tested by experiment whether it be true that two stones, one weighing ten times as much as the other, if allowed to fall, at the same instant, from a height of, say, 100 cubits, would so differ in speed that when the heavier had reached the ground, the other would not have fallen more than 10 cubits."*

Sagredo tried the problem experimentally with a result which contradicted Aristotle. Although experiments are most important in Salviati's eyes, at this point he offers something more convincing: *"But, even without further experiment, it is possible to prove clearly, by means of a short and conclusive argument, that a heavier body does not move more rapidly than a lighter one."* This is the sentence I referred to in the title of the paper. Let us imagine a large stone fixed to a smaller one, Salviati says, thus their aggregate mass is larger than that of the larger one, hence the two together would fall faster than either the faster large or the slower small one would do alone. That would mean that the slow small stone makes the fast stone faster? The slow would accelerate the fast? Galileo-Salviati refutes Aristotle by making use of a rational consideration which he regards as more convincing than some direct experiment. This quotation prompted me to write this paper.

Simplicio, however, does not give in easily. His arguments referring to daily observations do not ring hollow: *"I do not find it easy to believe that a bird-shot falls as swiftly as a cannon ball."* He seems to have performed some similar experiment, and Salviati presumes that if dropping a 100-pound and a 1-pound ball from a height of 100 cubit, the larger *"outstrips the smaller by two finger-breadths"*. But if Aristotle were right, it would be outstripped by 99 cubits!

Salviati acknowledges that the media: air, water or mercury influence free fall to differing extents. One has to consider also the shapes of the falling bodies; a thin gold foil almost floats in air. The two finger-breadth difference might be due to similar effects. But he is a strong adversary in the debate: *"But, Simplicio, I trust you will not follow the example of many others who divert the discussion from its main intent and fasten upon some statement of mine which lacks a hair's-breadth of the truth and, under this hair, hide the fault of another which is as big as a ship's cable"*.

At that point, important for the dramaturgy of the debate, where the author shows his self-portrait-protagonist to be greatly indignant, he finds an apposite example to show how the inquisitive mind must make a difference between basic phenomena and eventual, additional effects. This, however, is not always easy. That was the reason why Sagredo dismissed Salviati's otherwise correct observation and had recourse to rational argumentation instead of the analysis of the experiment. (By the way, Galileo is seen here also as an excellent author.)

The idea was not new; it was already formulated by Plato, e.g. in Book VII. of the Republic. "*These decorations in the heaven, since they are embroidered on a visible ceiling, may be believed to be the fairest and most precise of such things; but they fall far short of the true ones, of those movements in which the really fast and the really slow—in true number and in all the true figures—are moved with respect to one another and in their turn move what is contained in them. They, of course, must be grasped by argument and thought, not sight.*" And he expounded the idea in more detail: "*By the use of problems, as in geometry, we shall also pursue astronomy; and we shall let the things in the heaven go, if by really taking part in astronomy we are going to convert the prudence by nature in the soul from uselessness to usefulness.*" That is the question is formulated by observation or experiment the answer, however, is given through "*prudence by nature in the soul*", in present-day terms by rational thinking.

In one single sentence, Galileo formulated the basic principle of scientific research, considered to be true ever since. Experience must be revised by ratio; bare observations can never reveal the truth. That is the reason why the use of thought experiments is most often inevitable, the outcome of which does not depend on the eventuality of the experiments. *Max Planck* wrote: "*It is utterly senseless to state that a thought experiment is meaningful only then if it can be experimentally realized. […] The researcher's spirit is elevated above the world of real instruments by thought experiments, he is enabled to create new hypotheses, to formulate new questions which, if investigated by real experiments, might take him to new laws, to relationships which are inaccessible for any direct measurement. […] Obviously, a thought experiment is an abstraction. This abstraction, however, is as inevitable for the physicist, let he be either theoretician or experimentalist, as the fact that a real outer world does exist.*"

As far as the deviation of the results from the theoretical expectation was concerned, Galileo was not content with a reference to the chance variation of external conditions. The reason of the deviation must be revealed, and if not its magnitude at least its direction must be given, which means that theories must be conceived of how these conditions influence the results. One has to consider things like density and drag. Although the author is not always clear in telling apart the effects of the two, as to the gist of the approach he became the master of all posterity.

Salviati needs these arguments because Simplicio is still in doubt. "*I shall never believe that even in vacuum, if motion in such a place were possible, a lock of wool and a bit of lead can fall with the same velocity.*" (Frankly speaking we wouldn't believe it either, had it not been taught to us and would not have been demonstrated by experiment.) Salviati taking the objection seriously gives a very considerate answer, since he is unable to make experiments in vacuum—pumps are still not existent in his time. But "*if we find as a fact that the variation of speed among bodies of different specific gravities is less and less according as the medium becomes more and more yielding, and if finally in a medium of extreme tenuity, though not a perfect vacuum, we find that, in spite of great diversity of specific gravity, the difference in speed is very small and almost inappreciable, then we are justified in believing it highly probable, that in vacuum all bodies would fall with the same speed.*"

Salviati uses stepwise approximation and extrapolation as it has been done by any researcher since his time under similar conditions. He applies the method with honest caution calling the result only probable, not certain. Certainty must await direct measurements.

The sentence quoted and the relevant discussion is just a small gem from the cornucopia of the *Discorsi*. It is only Salviati's advice to Sagredo what I wish recall: "*Let the spirit of knowledgeable doubt with you!*"

And so with all of us!

Bibliography

Galileo Galilei. Dialogues concerning two new sciences, transl. Crew H, de Salvo A. New York: MacMillan Company; 1914.

Max Planck M. Lecture at the Harnack House, Berlin-Dahlem, 6. March, 1935. In: Physikalische Abhandlungen und Vorträge, Friedrich Vieweg und Sohn, Braunschweig; 1958.

The Republic of Plato, transl. Bloom A, 2nd ed. Basic Books; 1968, 1991.

30

The Birth of Chemistry from the Spirit of Doubt: Robert Boyle

Over the centuries, *Robert Boyle* has been held to be the founder father of modern chemistry. He was the first to realize that chemistry was not a sheer auxiliary of medicine or alchemy. With obvious pleasure, he stated that "*[f]or I observe, that of late chymistry begins, as indeed it deserves, to be cultivated by learned men who before despised it.*" One would expect to find unchallengeable statements, final words in the writings of a man who laid the foundations of a branch of science. This is not the case.

Boyle was born in 1627 into a wealthy English aristocratic family in Ireland. Having started his studies in Eton, he was sent with his private tutor on a long tour to Europe including a two-year stay in Geneva where he was greatly influenced by the city's air of Calvinism. Returning home, he started with his chemical experimentation first in Oxford (Fig. 30.1), later in London in his sister's house with the collaboration of several assistants, among whom *Robert Hooke* became particularly well known. Refusing all titles and ranks he was offered, he conducted a secluded life. He died in 1691 "*with so little uneasiness, that it was plain his light went out, merely for want of oil to maintain the flame*" as it was written by a contemporary. His library has been scattered, his grave is by now unknown.

The question of atomism—whether substances consist of a huge number of individual particles or are continual—had surfaced again in his years. Although he was greatly influenced by *Descartes*, Boyle's atomism stemmed from *Gassendi,* a philosopher who seems to have baptized the Greek atomist, *Epicurus.* His theses are of that kind: nothing can be formed out of nothing and there does not exist any substance which could be annihilated subject to the will of the Creator; atoms are in motion all the time because that was

© Springer Nature Switzerland AG 2019
R. Schiller, *Between One Culture*, https://doi.org/10.1007/978-3-030-20538-6_30

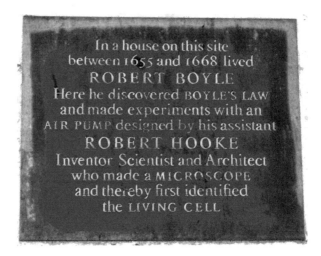

Fig. 30.1 Boyle's plaque in Oxford

meted out to them by God at their creation. This atomism, freed from athe-
ism, was greatly supported by some contemporary experiments. One of the
most important arguments against atomism, formulated already in Antiquity,
was that vacuum would exist between the atoms, whereas empty spaces,
regions without any matter, were simply not to be imagined. *Torricelli's* experiments with a tube closed on one end and filled with mercury demonstrated
that empty spaces do exist; this void has been called Torricellian vacuum ever
since. Thus, atoms can exist as domains completely filled with matter and
vacuum can exist as a region completely void of matter—no third possibility
can be conceived.

In that time, the development of science was greatly shaped by *Francis
Bacon's* inductive method. Disliking hypotheses and deductive statements,
Bacon believed in conclusions which are based on observations and their sys-
tematization. In accordance with that way of thinking, Boyle gathered a large
stock of empirical information performing a good many experiments himself
but also making use of every data, news, technical methods, opinions, guesses
which he could obtain from surgeons or artisans. Obviously, he made use of
the observations of the past but with criticism. The work of the alchemists and
that of *Paracelsus* were taken with a pinch of salt, comparing them to "*the
Navigators of Solomon's Tarshish fleet, who brought home [...] not only Gold and
Silver and Ivory, but Apes and Peacocks too*". A Baconian, he obviously recognized the importance of observations and experimenting but understood the
limits of such an approach: "*[...] yet I look upon the common operations and
practices of chymists almost I do on the letters of alphabet, without whose knowledge*

it is very hard for a man to become a philosopher, and yet that knowledge is very far from sufficient to make him one."

Boyle expounded his mechanical philosophy most cautiously. "*It seems not absurd to conceive that at the first Production of mixt Bodies, the Universal Matter whereof they among other Parts of the Universe consisted, was actually divided into little Particles of several sizes and shapes variously mov'd.*" The aim of his "*corpuscular philosophy*" was to perform experiments which make "*probable*" that nothing else is needed for the appreciation of "*almost*" all qualities (of a substance) but the "*bigness, shape, motion, and contrivance*" of the particles a substance is consisted of. "*Local Motion seems to be indeed the Principal amongst Second Causes, and the Grant Agent of all that happens in Nature*"—he wrote. (The prime cause is the Creator in the devout mind of a Christian.) The main points are not the indivisibility or persistence of atoms but the mechanical laws which control their motion. Mechanical philosophy in its strictest sense, atomism in a less strict sense.

He wrote about qualities paying little heed to the quantitative aspects of motion. Disliking mathematics, he thought that the phenomena of nature are unrelated to mathematical abstraction. (Notwithstanding this disposition of his it was he who formulated Boyle's Law, a mathematical expression for pressure and volume of gases, an equation much in use ever since.)

His atomism has been a most important step forward in science becoming a much cherished asset in our Intellectual Museum. His way of thinking, however, has not become antiquated. The title of his most popular book, *The Sceptical Chymist*, clearly reveals the author's frame of mind. Scepticism was not alien to his age. Philosophical scepticism was founded by the Greek thinker, *Pyrrho,* around the turn of the fourth-to-third century B.C. No written text of his in existence, what we know about the system is based on the writings of *Cicero* and *Sextus Empiricus*. The aim of the school, as of other Ancient philosophies, was to conduct a happy life . This can be attained, Pyrrho maintained, if we suspend our judgement. That was the precondition of the even frame of mind. "*Sceptical philosophy consists in the faculty to oppose experiences and ideas with one another by any means; as a consequence of this we attain the suspension of our judgement and hence the unperturbed mind.*" One has to realize that personal experiences and ideas are unreliable, whereas one's statements are influenced by tradition and prejudice. Thus any judgement is built upon shaky foundations.

The wording and phrases are most important in a sceptical text. Expressions like "*perhaps rather*" or "*nothing must be stated for sure*" are (must be) ubiquitous. Science is looked upon in a most curious way: "*We deal with scientific investigations because here we can always make a statement which opposes another one.*" (With all my affection to the school, I find it difficult to digest this sentence.) Scepticism is not a uniform theory; there are different degrees together with the most common counterargument against the system: whether scepticism itself is not to be considered sceptically.

The theory, or rather the way of thinking, was revived and welcomed in the age of Humanism and was given a high literary form by *Montaigne*, at first, through his essay on Raymond Sebond, even more through the style and general atmosphere of the *Essais*. His motto "*Que sais-je?*"—*What do I know?* is still with us.

Montaigne's *Essais*, published also in English a quarter of a century before Boyle was born, was the most popular book in England, second only to the Bible. Even in the absence of any direct evidence, it is difficult to believe that this book would not have had a marked effect on Boyle's way of thinking. In a sense, Montaigne is more coherent in his argumentation than classical stoics were. Pyrrho said that people who think they have found the truth deceive themselves. But, Montaigne adds, it is also sheer vanity to believe that truth cannot be revealed through human endeavour. You cannot even be sure whether you have failed. Montaigne had a poor opinion of science: scientists assume things that they themselves admit to be concocted—this angry sentence happens to refer to the Ptolemaic epicycles.

Boyle, of course, is neither uneasy nor angry with science. He doubts its statements. Not because he is uncertain of his own knowledge. On the contrary, doubting is his method of gathering knowledge. Considering the most important question, whether atoms are divisible or are not, he refrains from taking sides, his scepticism indicating clearly that further investigations are needed before the final word can be said. Because he thinks one may well be in doubt whenever one is not compelled by reasoning to make an assumption.

His caution in the exposition of mechanical philosophy, as quoted above serves, the same purpose: we are only at the beginning of the way, scepticism and doubt mark the further necessary steps. The book ends in the same vein.

"I shall not be so far in love with my disquieting doubts, as not to be content to change them […]. *And that as unsatisfyed as the past discourse may have made you think me with the doctrines of the Peripateticks, and the chymists, about the elements and principles, I can yet so little discover what to acquiesce in, that perchance the enquiries of others have scarce been more unsatisfactory to me, than my own have been to myself*".

Boyle and the philosopher *Spinoza* exchanged a good many letters through *Oldenburg*, the Secretary of the Royal Society. Once Oldenburg wrote to the philosopher "*our Boyle is not of the number of those who hold so fast to their own opinion that they do not need to take account of the agreement between them and the phenomena.*"

Boyle has taught science how to doubt.

Bibliography

Anstey PR. The philosophy of Robert Boyle. London: Routledge; 2000.

Boyle R. The sceptical chymist. London: J.M. Dent & Sons; 1911.

Hunter M, editor. Robert Boyle reconsidered. Cambridge: Cambridge University Press; 1994.

Laidler KJ. The world of physical chemistry. Oxford: Oxford University Press; 1993.

Montaigne M. The essays, Project Gutenberg, EBook #3600; 2006.

Partington JR. A history of chemistry, vol. II. London: MacMillan; 1961.

Saintsbury G. A short history of English literature. London: MacMillan; 1966.

Sextus Empiricus. Outlines of Pyrrhonism, transl. Bury RG. Cambridge, MA: Harvard University Press; 1933/2000.

31

Not Talking About Hume

You happy Greeks! Aristotle announced a firm belief at the very beginning of his Physics. "*For we do not think that we know a thing until we are acquainted with its primary causes or first principles.*" Coming across a phenomenon his question was: *why*. Millennia later, a most important turn was made by Galileo; from his time on the researcher has only been compelled to ask: *how*. As pointed out by historians, that change can be traced in Galileo's way of thinking and expressions. In his early years, considering motion, he wrote: "*since what we are looking for are the causes of the phenomena*". Salviati, the protagonist Galileo based on himself in his late book, *Dialogo*, is more cautious: "*At present it is the purpose of our Author merely to investigate and to demonstrate some of the properties of accelerated motion (whatever the cause of this acceleration may be).*" Newton's method is clearly based on observation and experience as well. His *Principia* gives answers to the questions *how* with unprecedented generality. It is systematic empiricism expressed in strict mathematics.

The ideal of a physical theory has been Newton's method ever since. The aim is to describe more and more observations by as few equations, called axioms, as possible. Understanding a phenomenon or a process has meant the recognition of the path which takes from the basic equations to the observations ever since. If a scientist announces having found the cause of a phenomenon, that means he could find its place and status in the mathematical formalism of generalized basic observations. It is most natural that this way of thinking, the stance of *hypotheses non fingo* (I do not make any hypotheses), was readily embraced by Hume, the thinker who paid attention only to phenomena dismissing the existence of any hidden causes.

© Springer Nature Switzerland AG 2019
R. Schiller, *Between One Culture*, https://doi.org/10.1007/978-3-030-20538-6_31

Having been invited to a conference held by the occasion of the bicentenary of Hume's birth, I was prompted to re-read Hume and tried to find out the echo of his extremely consequent empiricism in scientists' thinking. Sticking to my last, I kept to classical problems mainly in the area of physical chemistry.

I must admit to have become seriously disappointed. I could barely find his name in books on the history and philosophy of science, let alone technical texts. Although I did my best to find authors with opinions similar to Hume's. For example, Mach describes Hume's theory in a short paragraph, adding the phrase "*something we agree with*". Boltzmann cites his name only in connection with a not too edifying story. Wilhelm Ostwald, a consequent antagonist of atomic theory as long as he only could, wrote his book about the basic principles of chemistry almost in Hume's vein. "*A law of nature means that, according to experience, there exists a connection between two, definite and different observations which always appear simultaneously or in regular order. […] A law of nature is revealed as an expectation regarding the connection between possible observations. It follows that the notion of natural law does not imply any unconditional necessity or constraint it being based much more on the observation of the spatial and temporal connections of different events.*" Still, Hume's name is not even mentioned.

He might have a serious reason to keep silence. Discarding atomism, he embraced and developed the doctrine of energetics, a subtle theory which, being free of the clumsy, materialistic models of atomism, is based on the tenets of thermodynamics: the conservation of energy and the increase of entropy. Obviously, the laws of thermodynamics stem from observations; still they are difficult to treat in terms of strict empiricism. This is so because they do not refer to directly observable, measurable quantities. There does not exist any method or instrument by which one could measure energy or entropy. Was it Ostwald's intellectual honesty by which he was withheld from recalling Hume?

Hume is generally held to dismiss causality. The classical example: two clocks might chime at the same time and always at the same time without the sound of one of them being the cause of that of the other. A hundred years before Hume's time, the answer was already given by Guelincx: there does exist a common cause; it is the watchmaker who had adjusted the clocks. Hume is also held to dismiss determinism. Pullman writes in his book on the development of atomic theory: "*Foremost [among the most cherished beliefs of scientific study] was the conviction that all phenomena of nature are governed by strict determinism. Admittedly, a few philosophers, such as David Hume, had occasionally challenged the notion of causality.*" The two sentences taken apart are obviously true. Together, however, they create the impression that determinism and causality are synonyms. That is not true.

Every example, every description by Hume assumes that determinism prevails without any doubt. Hume maintains that a natural law is nothing else but custom, based on experience. This, however, is possible only because *"we assert that, after the constant conjunction of two—heat and flame, for instance, weight and solidity—we are determined by custom alone to expect the one from the appearance of the other."* The constancy of the conjunction is not doubted; on the contrary it is regarded as fundamental; otherwise, no experience-based custom could ever develop. The recurring example is that of the billiard balls. *"When I see, for instance, a billiard-ball moving in a straight line towards another; even suppose motion in the second ball should by accident be suggested to me, as the result of their contact or impulse; may I not conceive, that a hundred different events might as well follow from that cause? May not both these balls remain at absolute rest? May not the first ball return in a straight line, or leap off from the second in any line or direction? All these suppositions are consistent and conceivable. [...] In vain, therefore, should we pretend to determine any single event, or infer any cause or effect, without the assistance of observation and experience."* Anything can be imagined but only a single thing can be experienced.

One has to consider a figure like the one below (Fig. 31.1).

We experience one and the same process if the conditions are the same. In this example, if the first ball arrives at a certain angle and velocity, the second one will move as it did in any of the previous cases. If repeated a thousand times, we will observe the same a thousand times. There is no reason to imagine some cause here. But custom, based on experience, prevents us from having any doubt on how the events proceed. In different words, phenomena are deterministic without being causal according to Hume.

Let us now consider an even simpler billiard, something which Hume was unable to investigate simply because it was still unknown in his time. Here,

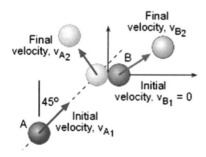

Fig. 31.1 Two encountering billiard balls with their velocity vectors

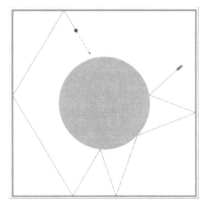

Fig. 31.2 The Sinai billiard

there exists only one single ball which dashes against the walls at the rims of the table and a cylinder standing in the middle of the table. The collisions are completely elastic and the ball rolls without any friction. The figure depicting this simple toy, called Sinai billiard, is given below (Fig. 31.2).

Simple as it is, its behaviour is curious because the path of the ball cannot be predicted, irrespective of its initial direction and velocity. There is no recurring experience to nurture some custom; thus, there is no custom to make any prediction on. One may observe the motion of the ball for periods of any lengths; still, one would never get the feeling that the direction of the rolling ball is predictable after the next collision. This motion cannot be regarded as deterministic in any traditional meaning of the word. But as far as cause and effect are concerned, one knows no more and no less than in the case of the traditional billiard.

Both billiards are causal in the Newtonian sense because the balls obey the same classical laws of motion in both cases. The difference lies in the actual conditions of the collisions, not in the governing laws. The laws rest on experiences which have been generalized in such a manner that any motion of any macroscopic body can be described by them. Also, of course, the billiard balls of the above two examples. Now it is mainly a matter of taste whether one denotes this axiomatic generalization of observations as empiric.

Today a good number of processes have become known which are described in terms of causal laws (using the traditional meaning of the word); still, their behaviour appears to be random, controlled by chance. These processes are causal but not deterministic; that is, they are just the opposite of Hume's description of nature. Their common name is *deterministic chaos*, a somewhat self-contradicting designation; perhaps causal chaos would be more appropriate.

Chaotic processes are often described in terms of probability calculus. While it is impossible to tell which one of the possible events will take place, the probability of their occurrence can be evaluated. Such a way of thinking might please Hume. According to his way of thinking, all our predictions are of a probabilistic nature since we only find probable that what was observed in the past will be observed also in the future. His natural philosophy makes use of probabilities only.

He, however, despite all his reverence towards mathematics, was no mathematician. In his vocabulary, probability is a subjective impression, a feeling and no well-defined quantity. Writing about probability, he apparently means high probability, a value near 1, something which is almost sure. Usually he does not consider the probabilities of events which everyday common sense would not expect to happen. Poisson saw a further problem in Hume's views. The probability of a statement is to be calculated as the number of favourable events in the past divided by the number of all the events in the future. But the number of future events is infinite; hence, this fraction becomes always equal zero prohibiting to draw any conclusion.

Still Hume, when talking about probabilities, might have conjectured an idea which was developed more than a century later. The world does not consist of pairs of billiard balls. Even the smallest body is built of an immense number of particles which influence each other's fate and motion in the most complicated manner. Their individual laws of motion are impossible to calculate, irrespective of the philosophical content of the word. The investigation must be limited to the determination of the fraction of the particles which behave in a certain way. For example, one can tell which fraction of the particles have a certain amount of energy without knowing the actual energies of each particle. In other words, one can determine the probability of a certain energy to occur in the multitude of particles. This is the basic question of statistical mechanics, a discipline born not from philosophical considerations but the internal development of mechanics and thermodynamics. It renders only probabilities, never certainties.

A central notion of statistical mechanics is called thermodynamic probability. Given the high number of particles, some average quantity, for example average energy, can be realized in many different ways: only few particles have very high, the rest have much lower energies, or all of them have almost the same energies, or … The number of possibilities by which a given average can be realized is called thermodynamic probability. A theorem of most general validity posits that if a system is in equilibrium, thermodynamic probability is at its maximum. Out of equilibrium, it is always smaller and it increases as the system gets equilibrated. This increase consists in the decrease of our knowledge about the system—obviously so, because if the system has more and more possibilities to choose, the observer knows less and less which one of them will be chosen. This increase of uncertainty is one of the most basic laws of nature. It is tantamount to the increase of entropy. Hume might have been happy, had he known that nature is controlled by a law of probability. But he would have been astonished understanding that it refers not to our existing but to our diminishing knowledge.

Anyway, what is the matter with Hume? Why is science history bashful to talk about him? I think even the most drab of data collecting scientists think more of his/her own results than what would have been ascribed to them by Hume. Because Hume bars science from the future, in two senses.

He has a frightening sentence: "*since it implies no contradiction that the course of nature may change, and that an object, seemingly like those which we have experienced, may be attended with different or contrary effects.*" As far as logical contradiction is concerned, that is completely true. But it makes any effort vain to recover natural phenomena which happened in the past or predict those which will happen in the future. Of course, he knows this only too well. "*If there be any suspicion that the course of nature may change, and that the past may be no rule for the future, all experience becomes useless, and can give rise to no inference or conclusion.*" All science, thus, becomes useless.

Take geology, for example. It may well be that the theorem by Lyell, the principle of actualism, the idea that the forces acting in the geological past are the same as those of the present day, is false. Maybe, but then geology ceases to exist. The geological situation brought about by these forces obviously changes with time. So does the information about the forces, together with the relevant models and mathematical descriptions. But all geologists are firmly convinced that the forces do remain the same and cause the same effects.

Being neither a historian nor a philosopher of science, I would never dare to walk in the forest of scientific revolutions, verification and falsification, tacit knowledge and pre-stabilized harmony. I share only the common understanding that measurements, data, theories change and evolve, earlier laws are superseded by more general and better founded ones, our knowledge increases and our way of thinking develops. But the phenomena are the same all the time. The lodestone was thought of differently by Plinius Maior and by Gilbert, again differently by Faraday and very differently by Heisenberg. The lodestone, however, attracted iron in exactly the same manner during all these centuries. Science makes sense only because we are convinced that "*the course of nature*" is not subjected to any variation.

Invariability is also the foundation of the predictive power which is regarded to be the greatest virtue of any theory. Hume seems to be ignorant about it. A good theory consists not only in the correct description of the observations for which it was originally meant but also for those which were not within scope when it was constructed. Perhaps even for those which were still unknown at the time when the theory was born. Dalton's naïve molecular models have led the chemists to an understanding of isomerism, a phenomenon unknown to Dalton. Maxwell's equations describe the propagation of radio waves, although Maxwell had no idea about their existence. Quantum mechanics was developed by Heisenberg and Schrödinger for the description

Fig. 31.3 Masaccio: The Expulsion from the Garden of Eden Cappella Brancacci, Santa Maria del Carmine, Florence

of the optical spectrum of atomic hydrogen and soon it became clear that the theory is appropriate to an understanding of the chemical bonds in a molecule. Such results, and many more, would be difficult to develop by an empiricism which is undecided on whether the observations stay firm as time goes by.

A well-constructed and thoroughly understood theory offers a broader view than the observations it is based upon. "*The theory takes further than the direct experiments by making use of the so-called thought experiments which are not exposed to the inadequacies of the real instruments*"—Max Planck wrote. A good number of important results are due to thought experiments based on the wise generalizations of the observations.

Take for example the case of Adam. "*It is evident that Adam, with all his science, would never have been able to demonstrate that the course of nature must continue uniformly the same, and that the future must be conformable to the past.*"—Hume says. This sentence is an archetype of a thought experiment. Most probably it is true since Adam had no previous experience of the course of events. Previously, he did not experience what the consequences are, if he eats of the forbidden fruit. As we all know the consequences were such as can be seen below (Fig. 31.3).

Bibliography

Aristotle. In: Barnes J, editor. Physics, The complete works of Aristotle, Vol I. Princeton University Press, Princeton; 1984.

Böhme. Die Wahrscheinlichkeitslehre bei David Hume. Berlin: Otte und Muhsoldt; 1909.

Boltzmann L. In: McGuinness B, editor. Theoretical physics and philosophical problems. Dordrecht: D. Reidel; 1974.

Chandrasekhar S. Newton's principia for the common reader. Oxford: Clarendon; 1995.

Ducheyne S. Galileo's interventionist notion of "cause". http://philsci-archive.pitt.edu/2580/

Faraday M. A speculation touching Electric Conduction and the Nature of Matter (1844). In: Classical scientific papers—chemistry (D.M. Knight arranged and introduced). London: Mills and Boon; 1968.

Hume D. Enquiry concerning human understanding. www.earlymoderntexts.com/assets/pdfs/hume1748.pdf

Hume D. Appendix. http://www.davemckay.co.uk/philosophy/hume/

Mach E. Die Principien der Wärmelehre. Leipzig: Barth; 1900.

Mach E. Erkenntnis und Irrtum. Leipzig: Barth; 1917.

Ostwald W. Prinzipien der Chemie. Leipzig: Akademische Verlagsgesellschaft; 1907.

Planck M. Die Physik im Kampf um die Weltanschauung. Vortrag, gehalten am 6. März 1935 im Harnack-Haus Berlin-Dahlem Johann Ambrosius Barth Berlin; Leipzig; 1935.

Pullman B. The atom in the history of human thought. Oxford: Oxford University Press; 1998.

32

Physicists' Molecule, Chemists' Atom?

"*…it is not chemistry that grows old, but chemists.*" That was what Mr. Boussingault, being in his late fifties, said at Karlsruhe, the September 3, 1860. The occasion was the chemical congress, the second ever international scientific congress[1], which was convened by Kekulé. He realized the pressing need for personal contacts and exchange of ideas in order to find a common conceptual basis for chemical thinking. The chemists' zeal of experimenting and the development of chemical industry rendered a large collection of phenomena and observations by the middle of the nineteenth century, demanding models and theories which enable the chemists to understand what they saw. The great success of the physical theories had set an example of the mathematical description of nature. However, a good many of the chemists refused the idea of applying the laws of physics to the phenomena of chemistry. The system of physical laws being rather a model than a tool they tried to develop the independent laws of chemistry.

It would be unfair to our chemist ancestors if we shook our heads at their desire for independence. They might know and understand the disciplines of motion and forces; the physics of their age, however, could have but little to say about the structure of substances. Mass attraction and electrostatics were known; there were even attempts to apply them to chemical phenomena, obviously with little success. Even Newton's mind was unable to unify mechanical laws with (al)chemical observations. Still, some chemists' views were different: for example, that of Lomonosov, who was the first to define physical chemistry as "*the science which explains by the laws and experiments of*

---

[1] The first ever international scientific congress was held at Szklenó—Sklené Teplice—Glasshütte (Austria-Hungary) on September 27, 1786 convened by the mineralogist and chemist Ignaz Born.

© Springer Nature Switzerland AG 2019
R. Schiller, *Between One Culture*, https://doi.org/10.1007/978-3-030-20538-6_32

physics the events taking place in composed bodies during the course of chemical operations". However, Lomonosov was as unknown as Newton, the alchemist.

Kekulé's aim seemed to be very modest. He wanted to achieve that the notions "atom", "molecule" or "equivalent" should carry the same meaning for every chemist and that both nomenclature and symbols should become uniform. This task necessarily involved some agreement as far as the structure of substances was concerned. Cautiously, the organizers maintained that whereas they do not expect all conflicting views to harmonize immediately they do hope "*that such works could pave the way for a much desired agreement between chemists in the future*".

Considering the dates of the years, one might think that all these problems had already been outdated around the middle of the nineteenth century. It was at the very beginning of the century when Dalton wrote that "*These observations have tacitly led to the conclusion which seems universally adopted, that all bodies of sensible magnitude, whether liquid or solid, are constituted of a vast number of extremely small particles, or atoms of matter bound together by a force of attraction...*" The atomic structure of the substances was commonly accepted by those years. Similarly, some years later, Thomson was of the same opinion: "*...it is the general opinion that [the bodies] consist of atoms, or minute solids, incapable of further division*". He was cautious, nevertheless: "*With respect to the nature of the ultimate elements of bodies, we have no means of obtaining accurate information...*" Dalton did not revive the idea of the indivisible atom; as a matter of fact, it was always alive. Also, it was known that compounds are composed of elements. What he and his contemporaries, first of them Gay-Lussac and Avogadro, did was that they *assumed* that the smallest particle of any compound, called a molecule, is composed of a definite and invariable number of atoms of the elements. The molecules of a compound were regarded as perfectly similar to each other.

The weights of substances, the volumes and pressures of gases and the temperature—that was all what the chemists of Dalton's age could measure. With such a meagre source of information at hand, they succeeded in establishing the elementary composition of compounds, called empirical formulae. The development was, of course, not quite smooth; researchers often misunderstood each other's arguments; nevertheless, the idea of the molecule composed

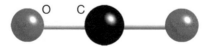

Fig. 32.1 Ball and stick model of carbon dioxide

of a well-defined number of atoms started to prevail. Models made from balls which represent atoms and sticks between the balls the latter symbolizing chemical bonds were easy to imagine, construct and prepare in the workshop (Fig. 32.1).

Most chemists seemed to be happy with this level of understanding, which also enabled them to model chemical reactions by imagining balls getting loose, some of them to rearrange and new sticks to be fixed. Nevertheless, they often disagreed as far as empirical formulae or chemical symbols were concerned; hence, the most direct aim of the Karlsruhe Congress was to eliminate or at least mollify such disagreements. Canizzaro, an Italian chemist, was the most effective in fulfilling this task. Critically summarizing all what was measured, calculated and inferred in the course of about half a century before the meeting, he was able to draw a non-contradictory picture on composition of molecules, valency, molar weight and molar volume, notions which have been thought to be sound and valid ever since. It took some time for the participants to digest what he said but in a relatively short period of time Canizzaro's views gained ground. That is what is taught as the basics of chemistry even nowadays.

There was, however, a deeper-lying problem. The word "assumed" was italicized above, there being no direct evidence of the existence of atoms and molecules. It was only a hypothesis. Critically thinking chemists, or physicists, who knew very well how to prove experimentally a theoretical statement, were not too eager to accept the idea. The graphic character of the chemists' molecule, the tangible wooden model, was against their lofty ideals of a theory.

J.C. Brodie, an Oxford professor of chemistry, having seen a ball and stick model felt nothing else but disgust calling it a "*thoroughly materialistic bit of joiner work*". (Materialism was a word of abuse in that age.) Instead, he proposed a system of algebraic operations based on nothing else but the relative weights of the elements and compounds, a method which soon was proved to be both erroneous and useless.

Even Faraday wrote that he had "*grown out*" the idea of the atom thinking that Dalton's theory was a most clumsy hypothesis. He preferred Boskovic's proposition on matter built of elementary, point-like centres of force. Developing this hypothesis, he thought that the formation of a compound consists in the unification of independent centres similar to the interfering waves of water. Was this a conjecture by a genius of the wave mechanics of molecules discovered almost a century later?

Maxwell, a convinced atomist, was one of the founders of statistical methods by which one can evaluate the aggregate behaviour of a substance (originally of a gas) by knowing the properties of its constituent molecules. But he was cautious in his inferences: "*It is not essential, however, to the mathematical investigations to assume that the molecule is made up of atoms.*" Kekulé did not disagree with this sort of reservation: "*Thus it is not established if the smallest quantity of a substance that enters into a reaction is also the smallest quantity of this substance that plays a role in heat phenomena.*" Canizzaro did not share the doubts of these great men, feeling certain that atoms and molecules are the same entities in physics and chemistry. Sure, this stance was conceptually less demanding than those of the adversaries and it was only partly susbstantiated through experiments. However, as it happened, the joiner work of Dalton, Gay-Lussac and Avogadro was seen to be in agreement with both the phenomena of chemistry and with the statistical theory of the substances. The same molecular models had to be made use of in both disciplines.

Still, the basic assumption had to await several decades to get proven. The atomic-molecular model was seen to be effective, useful and practicable, but the existence of the atoms was still not supported by any direct experiement. If a gas consists of a great number of haphazardly flying particles of different velocities, then their density must vary from point to point as chance makes them gather at a certain site and disperse at another one for a fleeting moment. This phenomenon, called density fluctuation, theoretically decribed by Einstein, was proven experimentally; the blue colour of the sky is due to this process.

This was a serious argument in favour of atomism, which was also supported by another idea of Einstein's. If the molecules are entities as real as larger particles they must move, walk and diffuse according to similar laws. The experiments of Perrin demonstrated this similarity. Hence, by the beginning of the previous century, any doubt referring to the real existence of atoms and molecules disappeared. Molecules consist of atoms. An atom is the smallest particle of an element which cannot be altered by any chemical or physical process. It is indivisible and unchangeable.

Is it? It happened about a decade *before* the atomic-molecular structure of the substances was proven that radioactivity was discovered, a process which consists in the decomposition of atoms showing that they can change and divide. First, it was seen to proceed spontaneously; later it could be effected by external means in particle accelerators and nuclear reactors. The notion of the atom has lost the solidity given to it by chemistry.

We have been sure for more than 150 years that a molecule is the same entity in physics and chemistry. We have been sure for more than 100 years that an atom is not the same entity in physics and chemistry.

Bibliography

Brock WH, editor. The atomic debate—Brodie and the rejection of the atomic theory. Leicester: Leicester University Press; 1967.

Cannizzaro S. Sketch on a course of chemical philosophy. In: Nye MJ, editor. The question of the atom. Los Angeles: Tomash Publishers; 1984.

Faraday M. A speculation touching Electric Conduction and the Nature of Matter (1844) In: Classical scientific papers—chemistry (D.M. Knight arranged and introduced). London: Mills and Boon; 1968.

Maxwell JC. On the dynamical evidence of the molecular constitution of bodies (1875). In: Nye MJ, editor. The question of the atom. Tomash Publishers, Los Angeles; 1984.

Partington JR. A history of chemistry, Vols. III, IV. London: MacMillan; 1962, 1964.

Wurtz C-A. Account of the Sessions of the International Congress of Chemists in Karlsruhe, on 3, 4, and 5 September 1860. In: Nye MJ, editor. The question of the atom. Los Angeles: Tomash Publishers; 1984. Selected Classical Papers from the History of Chemistry (Ed. Carmen Giunta). http://web.lemoyne.edu/~giunta/papers.html

33

Why Is Ortega Angry with the Scientists?

He is angry with the scientists not with science. Better to say he is not angry with them; he rather despises and looks down on them. He despises them as he does the masses in general and is in awe of them as of the masses in general. But perhaps his feelings are not the main issue. More important seems to be the realization of the sociological fact that "*the present day scientist is the prototype of the mass-man*". This is stated in his magisterial essay book which both by its title and content reports about *The Revolt of the Masses* as the terrifying but inescapable development of his age. In this context, the above-quoted sentence is definitely not flattering and its continuation is even less so: "*And this has not happened just by chance, nor as a consequence of the scientist's personal weakness, but he is made a mass-man automatically by science itself, which is the root of our civilization, so he becomes barbarous and primitive.*"

Ortega analyses in detail the history of the process which degraded the scientist. The reason of the intellectual degeneration is seen in the specialization of the different branches of science. In earlier days, a man of learning was adept in all walks of science. The present-day hero and idol is the specialist of his own narrow and ever narrowing field. The person "*who is bold enough to consider as a virtue of his own ignorance in any field which lies beyond his narrow speciality and calls* dilettantism *the aspiration for some general knowledge*". Ortega is amazed that these (in all senses of the word) limited specialists are still able to develop science and not only in their own field but also in basic far-reaching contexts. This is so, he thinks, because people learn the technicalities of science, or rather because there exists something like the techniques of research which can be taught and learnt. Hence, the whole field "*has become*

© Springer Nature Switzerland AG 2019
R. Schiller, *Between One Culture*, https://doi.org/10.1007/978-3-030-20538-6_33

perfunctory", a fact which enabled *"astonishingly mediocre people"* to produce outstanding results.

Ortega, of course, knows only too well that not every scientist is of that type. He has sincere reverence for the real scientist. Talking about the birth of experimental sciences, he writes about *Galileo* and *Newton* repeatedly and, when speaking about his own age, he often thinks of *Einstein*.

One might meditate on whether Newton or Einstein were encyclopaedic minds or limited specialists; were such a meditation not almost sacrilegious. The question, however, is very different. To be a mass man is not just a condition but it is a sin and a danger, Ortega thinks. This was not always so, he says. In our days, however, … *"If the mass is about to act on its own then it revolts against its own fate; it is this what is going on now and that is the reason why I am talking about the revolt of masses."* And that is the way how the specialist is about to revolt, although being a cobbler stuck to his last he still tries to grow beyond it. Although ignorant in every field other than his own he is *"supercilious in areas unknown to him"*.

This analysis, based obviously on a number of shrewd and wise observations, is somewhat odd: at the same time, Ortega has a deep respect for both the authentic scientist and the technician. *"The man of science has become the pariah of the society."*—he writes in 1930. (In the context of the text, this appears as a compliment.) Further, he states that the formula for the nineteenth century was liberalism+technique, whereas for the twentieth century it is capitalism+science. Clearly, he understands that the enormous development which started in the eighteenth century is rooted in science and only contemporary European technique has a scientific basis. That has been the reason why the East has lost its superiority in technical matters.

Even so, it is intolerably and apparently inevitably dangerous that mediocre people who know very much about very little strive to attain influence and power in society.

It is a mean business to take the prophet by his word ninety years later—years which were pretty rich in experience. To be wiser now than he was then is of little merit to us. We do know that there existed a moment in modern history when science had become from an indirect and unwitting tool a most conscious shaper of politics and war. From this moment on, scientists could and had to make decisions in world politics. Partly because, probably, some of them wanted to do so. But partly because only they had the knowledge which was indispensable for any decision. These people were not the horde of well-trained technicians, but the most brilliant brains of the age who could nurture exceptional ideas. That was something that Ortega was unable to foretell. It was the revolt not of the limited-thinking mass but of the aristocracy of mind and spirit. Not of the workers but of the kings.

The story of course continued. This particular moment prompted a development of sciences which was never seen before either in width or in depth and—sure—also in specialization. These events may justify Ortega's disdainful pessimism.

Or rather not. There is little sign that intellectuals in science or technology would aim for power in the society. These people have enjoyed their advantageous position and tried to make use of the standing their knowledge provides for their own work and for their own benefit. But little was heard about some collective move or personal ambition to enlarge the lab in order to engulf the whole society.

Perhaps one could talk about us, limited specialists, in a more polite tone. Heisenberg compared us to the medieval stonemasons. It was not the mason who created the great ideas of religion. Perhaps he was unable to understand much of them. But when thoroughly carving the folding of a saint's robe or an angel's feather, he served the same idea, a glimmer of the realization of this idea, as the tortured martyr or the meditating theologian.

Is this an optimist's exaggeration? "*He who does not want to exaggerate is bound to keep silence; moreover he must switch off his mind and find the way how to grow dull.*" This is also something I have learnt from Ortega.

Bibliography

Ortega Y, Gasset J. The revolt of the masses. New Ed. New York: W. W. Norton; 1994.

34

The Sage and the Ditch

Let us begin with an ancient anecdote handed down over centuries and millennia. It is about the first philosopher of Hellas, the great mathematician, Thales of Miletus. It was retold also in the sixteenth century in England. "[Thales] *fell plumpe into a ditche over the eares. Wherefore an olde woman that he kepte in his house laughed and sayde to him in derision: O Thales, how shuldest thou have knowlege in hevenly things above, and knowest nat what is here benethe under thy feet?*" (Fig. 34.1).

That is the traditional image of the scholar. The abstract mind treading in the dust of the day fixes his eyes in the sky, and it is exactly his extraordinary mind that impedes him from being well versed in worldly matters. Our sense of justice seems to be inexhaustible, refusing the idea that one person would obtain everything, whereas his neighbour would get just nothing. The wise man is poor and helpless, and the rough bumpkin is successful and rich. "*Without Learning Or Education, He Left A Million Of Money Behind Him.*" These are the words on the self-composed epitaph of Trimalchio, the protagonist of the Satyricon by Petronius. Most probably, Trimalchio was grateful for his poor education.

However, the old woman's derision is perhaps not completely justifiable. There is another story, about Thales again, with a different moral. "*Thales, so the story goes, because of his poverty was taunted with the uselessness of philosophy; but from his knowledge of astronomy he had observed while it was still winter that there was going to be a large crop of olives, so he raised a small sum of money and paid round deposits for the whole of the olive-presses in Miletus and Chios, which he hired at a low rent as nobody was running him up; and when the season arrived, there was a sudden demand for a number of presses at the same time, and*

© Springer Nature Switzerland AG 2019
R. Schiller, *Between One Culture*, https://doi.org/10.1007/978-3-030-20538-6_34

Fig. 34.1 French illustration, eighteenth century

by letting them out on what terms he liked he realized a large sum of money, so proving that it is easy for philosophers to be rich if they choose, but this is not what they care about."

The gist of this story (improbable as it might be in the light of our experiences with meteorology) is not the abstract mind's impractical stupidity. On the contrary, the abstract mind brings in a good profit, measurable on the scales of the moneychanger. The story expresses an ethical opinion. The wise man is in possession of those mental faculties which could make him successful in money matters. However, his spiritual demands preclude his mingling into such affaires.

This is so much true that one can express one's sympathy for a scholar in terms of their feelings with wealth: not necessarily by the amount of money the scholar possesses, rather by what he is coveting for. In short, if the author likes a wise person, then he states that this person dislikes money.

Xenophon, for example, liked Socrates, accordingly he stated that he never accepted money from his students. "[…] *by this abstinence, he believed, he was best consulting his own freedom*". Socrates, however, disliked the sophists so he had a somewhat pragmatic definition of them: *"And in the same way we call those who sell their wisdom to anyone who wants it sophists, just as if they were prostitutes.*" He did not disclaim their wisdom but they ask money in exchange of their wisdom. Bad enough!

On the other hand, some people disliked Socrates, one of them being Aristophanes at least when he wrote his play *The clouds*. The comedy is mainly about Socrates asking for lump sums in return for his teaching.

Curious how this opinion about the lofty spirit feeling alien to the desire for wealth has been handed down from Antiquity almost until present. Perhaps one could ask why it is only the scholar whose greed is under anathema. The artist might live in luxury, may ask a fortune for his creations, scandalizing neither the everyman nor the connoisseur, neither in the distant past nor nowadays. On the contrary, the rank and talent of an artist have been measured in terms of guineas. Why? Because the artist is thought to be an extrovert, so his greed is readily accepted even expected by society, whereas the scholar, traditionally typecast as an introvert, plays his role better by exercising financial asceticism? Whereas in any other respect, both genera are regarded as the aristocrats of mind and spirit.

Whatever the reason for this belief or prejudice might be, it has been stubborn enough to remain vivid even in the age of the applied and engineering sciences, although practical success is known to be appreciated in terms of money. The scientist, however, was expected to disdain money. Faraday's bon-mot about his results, which would soon become assessable properties, might have been just a witty remark, notwithstanding the fact that our monthly electricity bill proves it to have had a taste of divination. Anyway, it referred to state revenues and not to the scientist's personal income.

Still, at the beginning of the nineteenth century, as practical tasks were cropping up and financially useful results were achieved, the class of scientists was undergoing some differentiation. Undergraduates at the German universities became divided into two groups, one of them being motivated by spontaneous intellectual pursuit of knowledge, the other by the hope of becoming a breadwinner. Young men (female undergraduates being non-existent) knew which class they belonged to right at the start of their studies. Mainly the social standing and the financial resources of the parents decided whether the freshman would pursue *Berufstudien* (learning by vocation) or *Brotstudien* (learning for bread).

Friedrich Schiller, delivering his first lecture at the University of Jena, made a sharp distinction between the two types of learning, unequivocally praising Berufstudien. He thought only people whose minds have been trained in that manner would be able to produce new ideas and achieve new results. The poor ones who attended the courses as part of their Brotstudien would always remain in the secure realm of what was well known and never doubted. These noble ideas were framed in the year of the French Revolution. A few decades later, they might have sounded somewhat outdated.

One of the first heroes of Brotstudien was the great German chemist, Justus Liebig. Son of a poor village pharmacist, he thought it to be only too obvious that research means work and work must be paid for. He must have been a resolute character. It turned out several years ago that his doctoral degree, indispensable as it was for a university career, had simply been bought by him. Not that he would have been unable to write a thesis. But it was expected to be composed and debated in Latin and he, being feeble in that language, did not want to squander his time. Soon he became a great practical chemist creating the idea and methods of fertilization, to mention just one of his important achievements. A neat slice of bread lavishly smeared with butter was his return by becoming university professor, millionaire, minister and baron.

The Hungarian industrial chemist W. Wartha being almost two generations junior to Liebig behaved in a way that must have been much more in accordance with the lay people's expectations. Being the founder of industrial chemistry in his country, he developed the technology of a glaze, called eosine, for the Zsolnay Factory at Pécs. While he did not attach his name to the industrial process and probably did not obtain any particular compensation for it, he was keen to deposit the description of the method with the Academy of Sciences. But that's fine! An ambitious mind's modesty in money matters.

The research which can be sold and the scientist who can be purchased degraded science in the eyes of the many. This caused also the loss of the scientist's personal authority. The emblem of the legend about twentieth-century science, Einstein's face, whose hagiography includes his seraphic aversion to money making, tries to restore the traditional figure. As Infeld wrote: "*Often he mentioned he would be willing to do some physical work if it were useful, for example working as a shoemaker, obviously without interrupting his scientific work. However, he declined to use his knowledge for making money for example by giving university lectures. […] Physics is a thing much too great and important to be exchanged for cash.*"

A difficult inheritance for the present-day scientist.

Bibliography

Aristotle. *The politics* (Transl. Intr. Carnes Lord). Chicago: University of Chicago Press; 1984.

English jest book Merry tales and quick answers (1530) https://en.wikipedia.org/wiki/The_Astrologer_who_Fell_into_a_Well

Infeld L. Albert Einstein his work and its influence on our work. New York: Charles Scribner's Sons; 1950.

Petronius. Satyricon (Transl. Allinson AR (1930) Modified and annotated Chinn C (2006)). http://pages.pomona.edu/~cmc24747/sources/pet_satyr.htm

Schillers Werke in fünf Bänden. Berlin: Aufbau-Verlag; 1984.

Xenophon. Memorabilia (Transl. Bonnette AL). Ithaca: Cornell University Press; 1994.

35

Why Be Reason-Able? Jung and Pauli Thinking Together

"*The discovery of discontinuity, exemplified by the energy quantum or the radio-activity, means the end of causality and consequently of the trinity of principles.*" This was the opinion of Carl Gustav Jung, the psychologist, decades after the birth of modern quantum physics. Here, the trinity of principles means space, time and causality, ideas which have governed classical physics, according to Jung.

His belief that these ideas are superseded by modern physics was based, among other sources, on a sentence of James Jeans, a great astronomer and physicist in the first part of the previous century. Jeans stated that "*radioactive break-up appeared to be an effect without a cause, and suggested that the ultimate laws of nature were not even causal.*" A sentence most easily misunderstood. As it is obvious from Jeans' other texts, what he meant was the inadequacy of classical one-to-one correspondence between cause and effect. Understanding the post-classical development of physics, he was compelled to see how chance and probability were coming to the fore. Instead of constructing mechanical models, theoreticians must make use of mathematics as their primary tool. "*From the intrinsic evidence of his creation, the Great Architect of the Universe now begins to appear as a pure mathematician.*"—wrote Jeans. There is little doubt that Jung miscomprehended what he read about the deficiency of causality.

Jung, however, did not waste too much time on theoretical physics. One of the basic notions of his work is the *collective unconscious*. This is made of instincts and *archetypes;* these latter being collectively inherited psychical structures which remain forever unconscious for any individual.

© Springer Nature Switzerland AG 2019
R. Schiller, *Between One Culture*, https://doi.org/10.1007/978-3-030-20538-6_35

Having get rid of the heavy load of causality, the psychologist, with a great inclination to transcendental thinking and mysticism, introduced the idea of *acausal phenomena.* Jung defined acausal correspondence as a temporal coincidence of events which seem to be causally unrelated and their chance coincidence looks most improbable. (At least for the mathematically untrained. Mathematicians have shown more than once that naïve expectations about chance are often misleading. People think it is miraculous to get the same seat in the theatre twice although this is quite probable. But they play lottery in the hope of winning a hoard of money, which is most improbable.)

Jung regarded acausal phenomena to be very common. In order to find a place for acausality in his edifice of psychology, he extended the idea of collective unconscious to the whole world; the collective unconscious permeating and unifying everything: human beings, animate and inanimate nature. It is the collective unconscious that is revealed through acausal correspondences, events which cannot be accounted for by rational considerations.

Jung was very effective in observing and collecting such events. His collection contained special cases of ESP (*e*xtra*s*ensory *p*erception): coincidences of words; the unexpected appearance of animals, for example a goldsmith-beetle in a room where its distant relative, the Egyptian scarab, happened to be mentioned; or the gathering of birds at a house where later someone was to die— just to mention a few of his items. His most thorough research, however, referred to the horoscopes of married couples where he maintained to have observed significant correlations between the stellar constellations of the partners. I am not going to copy these huge tables; besides, I do not know if these marriages turned out to be happy.

One might be astonished that Jung's irrational way of thinking and his anti-scientific views were nurtured in a period of his life when he closely cooperated with the great theoretical physicist, Nobel laureate, Wolfgang Pauli. The Paracelsian psychologist who usually argued with analogies, myths and dreams, and the acclaimed theoretician of strict science maintained a fruitful intellectual companionship.

Their common work started when Pauli, in need of psychoanalytic treatment, approached the noted psychologist. As a part of the treatment and under Jung's influence, Pauli had a long series of connected dreams. Jung was most happy, saying that "*the majority of dreams are usually unimportant, that is becomes useless with the eventual change of the external conditions. The important dreams, however, like yours, are of great use to the scientific research of motives.*" Finally, Pauli handed over the description of four hundred dreams to Jung, who asked for permission to publish and comment upon this material, obviously after deleting everything which could be a clue to the dreaming person's identity or to any of his personal affairs. This became the basis of the great essay entitled *Traumsymbole des Individuationsprozesses* (The Symbols of Dreams in the Process of Individuation), which is now a chapter of the opus magnum, *Psychologie und Alchemie*. One would expect that the scientific training and world view of the dreamer was useful to Jung since alchemy in some sense is still related to science. This was not the case. The fact that Pauli was a scientist was important only because this was the guarantee of his ignorance in history, philology, archaeology and ethnography, in one word of everything by which the dreamer could unwittingly influence the dreams. Jung regarded this as a reassuring proof of the exactitude of his psychological experiment: Pauli was pretty much uneducated.

The correspondence between Jung and Pauli lasted for decades. First, the dreams must have been discussed since they were often related to notions and signs which only a physicist could understand. Explanations were needed; hence, Pauli prepared a "vocabulary" for Jung about the "background physics" of the dreams.

"*Physical notions as archetypical symbols*" that is how Pauli's essay starts. The author makes clear that it is not about the analysis of a single person. Pauli wanted to achieve "*a description of nature which is based on a unified view of physis and psyche*". This can be done because "*I see physics and psychology as areas of research which mutually complement each other*". That makes dream symbolism necessary and that is why physical notions are to be treated as archetypical symbols. This view is supported by the personality and theoretical work of Kepler. Pauli thinks that "*[Kepler's] ideas obviously represent the transition from the old magic-symbolic to the modern quantitative-mathematical description of nature*".

Let the electric dipole be taken as an example. Pauli was happy to dream about it. The dipole is a small entity, a molecule, in which positive and negative charges are separated.

The physicist knows that if an external electric field acts on a multitude of dipoles, they get aligned in the direction of the field, an order which becomes marred with increasing temperature. Pauli understands his dreams about ordered multitudes as a representation of "*participation mystique*", a favourite notion of Jung meaning the multiplication of the individual spirit, the formation of a "*bush of spirits*". Disordering due to increasing temperature corresponds to the disruption of participation mystique as a consequence of individual differentiation. One must accept this interpretation; the physicist-dreamer certainly knows better.

Pauli had frequent dreams about radioactive nuclei, isotope separation, the splitting of spectral lines. In his vocabulary, the atomic nucleus is the symbol of the "*Selbst*", the Jungian *deep-self*. Whereas as far as the separated nuclei or the split spectral lines are concerned, Pauli makes use of his neither-physical-nor-psychological "*neutral language*" (this is his own expression). Splitting or separation refers to the division of something which is very important but can be realized only through most complicated methods. In physics, this might be an isotope separator or a spectroscope. In psychology, it is methodological imagination. His dreams about splitting and separation refer to the division of the psyche's content, the process at the end of which the unconscious becomes conscious.

Complementarity is an important notion in modern physics. Introduced by Niels Bohr, it means that some phenomena can be described and understood only in terms of two different and contradictory approaches. The most common example is light: it behaves both as wave and particle. Pauli thinks that in psychology such complementarity appears in the duality of the conscious and unconscious.

Beyond physics, the theoretician often had dreams about mathematical expressions. At the beginning of a dream, he saw a bird laying an egg which got divided all by itself. Later, the dreamer realized that he also had an egg is his hand, which was also divided—now there were four eggs on the scene. Four mathematical expressions appeared (two sine and two cosine functions) each belonging to one of the eggs. The four expressions got united (in terms of the Euler expression for complex harmonics, the mathematics was rigorously dreamed), the result being the equation of a circle. And indeed the last picture the dreamer saw was a circle.

In Pauli's interpretation, the dream represented the division of the content of the psyche, but in this case, it was followed by a unification process resulting in the appearance of the circle as the symbol of completeness.

Symmetry is one of the most important ideas in contemporary physics. To Pauli's mind, his seminal results are closely connected to it. Years after the interpretations of his own dreams, working on an important and difficult theoretical problem, he wrote a letter to Heisenberg with long calculations and strict physical considerations. The letter ended with a sentence, not unrelated to the theory: "*Division and reduction of symmetry, this then the kernel of the brute!*"

Theoretical physics or psychoanalysis? Physis or psyche? Perhaps Pauli could overcome the dichotomy.

Bibliography

Heisenberg W. Physics and beyond (transl. Pomerans AJ). New York: Harper and Row; (1971).

Jung CG, Pauli W. Naturerklärung und Psyche, Rascher, Zürich C.G. Jung, Synchronizität als ein Prinzip akausaler Zusammenhänge. W. Pauli, Der Einfluss archetypischer Vorstellungen auf die Bildung naturwissenschaftlicher Theorien bei Kepler; 1952.

Pauli W, Jung CG. In: Meier CA, Hrsg. Ein Briefwechsel 1932–1958. Berlin: Springer; 1992.

36

The Chemists' Not Too Old Glory

"*I had a walk with Thomas Mann on a beach some fifteen or twenty miles southwest of Los Angeles*"—this is the first passage of an essay by Aldous Huxley. Name-dropping, one might say, since the great man plays no further role. The text is about hygiene, both private and public. A walk on the seaside served as an appropriate introduction: accidentally they got near a city refuse dump, where they were startled by the sight of millions and millions of used contraceptives. The stink repelled them only a minute after. This experience induced a chain of thoughts on cultural history, theology and sociology in the head of the knowledgeable author.

He writes about long periods of human history when something that had been known millennia earlier at a number of places, including Crete in the Minoan times, was forgotten. It was the canalization of towns. He discusses the culture of baths in Antiquity and the Eastern habits of hygiene which became known during the Crusades and went into oblivion later. Also, the late start and slow development of soap manufactures in Marseille.

People never liked stink, filth and grimy bodies. However, the ethics of the Middle Ages regarded as unnecessary or sinful to like anyone's body, let alone one's own. That age did not want to pamper the human body, not even with a bath. Huxley could find a large collection of texts which proved how bad odour and repelling appearance assisted self-restraint. Strict morals went together with the virtue of modesty: it was thought to be indecent to get washed below the belt.

In the poor quarter of London's East End, waste water was dumped into puddles without any outlet even in Queen Victoria's time. The waste water of

© Springer Nature Switzerland AG 2019
R. Schiller, *Between One Culture*, https://doi.org/10.1007/978-3-030-20538-6_36

the canalized quarters was waving in the Thames. And no inattentive walker in the streets could tell what and when was going to be poured on his head.

The sad state of the huge city's river was reported by a committee in 1836. "*The Thames receives the excrementitious matter from nearly a million and a half of human beings; the washing of their foul linen; the filth and refuse of many hundred manufactories; the offal and decomposing vegetable substances from the markets; the foul and gory liquid from slaughter-houses; and the purulent abominations from hospitals and dissecting rooms, too disgusting to detail. Thus that most noble river, which has been given us by Providence for our health, recreation and beneficial use, is converted into the Common sewer of London, and the sickening mixture it contains is daily pumped up into the water for the inhabitants of the most civilized capital of Europe.*" Something needed to be done but this did not happen very soon. Huxley apparently did not know Faraday's letter written to the Editor of *The Times* some twenty years later. "*The appearance and the smell of the water forced themselves at once on my attention. The whole of the river was an opaque pale brown liquid. In order to test the degree of opacity, I tore up some white cards into pieces, moistened them so as to make them sink easily below the surface, and then dropped some of the pieces into the water at every pier the boat came to: before they had sink an inch below the surface they were indistinguishable [...].*" No doubt, the great experimenter performed also this observation according to his own standards. Finally, he drew the conclusion: "*If we neglect the subject, we cannot expect to do so with impunity; nor ought we to be surprised if, ere many years are over, a hot season give us sad proof of the folly of our carelessness*" (Fig. 36.1).

The imminent and recurring danger of cholera was a valid argument in favour of public hygiene. This, however, was interpreted in a small-minded way. First, canalization and waste water treatment were realised only in the parts of the West End inhabited by higher classes. Sometime later, it must have been realized that no comma bacillus honours social barriers. It was no earlier than 1896 that sanitary conditions became acceptable. As it is well-known, this was not the end of the story. By the second half of the twentieth century, the Thames became seriously polluted again and very expensive new treatments were needed in order to make water pure enough to enable trout to be caught. I hope the anglers of London have had nothing to complain about ever since.

Huxley's main concern, however, is not public hygiene but social inequality. A thing which can be measured also by soap consumption. He renders a good many quotations on how gentle folk took the poorer classes as a filthy, unkempt and stinking lot just because they were somewhat less filthy or could afford to conceal their unkempt bodies under expensive garments. Disagreeable

Fig. 36.1 A cartoon about Faraday's letter to the Editor of The Times

odour might devalue even a democratic procedure. Shakespeare's Coriolanus would never "*Appear i' the market place [...] To the people, beg their stinking breaths.*" Even death and life may hinge upon smell. In order not to be taken to Caesar's triumphal march, Cleopatra commits suicide, because she loathes crowds with their "*thick breaths, / Rank of gross diet, shall be enclouded / And forced to drink their vapour.*"

That is just literature. However, political argumentation might also hinge on statements about officials being persons elected "*by stinking breath*" and who "*stand upon the breath of garlic eaters*". Even good-hearted Victorian reformer philanthropists had to concede that "*by no prudence of their part can the poor avoid the dreadful evil of their surroundings*".

Historical experience taught Huxley that this type of social injustice was eliminated neither by social movements nor by revolutions and wars, but by chemistry. Chemical cleaning is somewhat younger than the railway and older than the transatlantic cable. Thanks to this technique everyone may wear clean dresses even without having new or fashionable ones. Poverty is not associated with dirt, dirt with untouchability, untouchability with contempt. Social classes are not segregated by disgust anymore.

It has since been rendered impossible to recognize the social status of a passer-by just by sheer look. This might have been the most important step in abolishing the caste structure of English society, the author muses.

Echoing the optimism of the mid-twentieth century, Huxley is proud to list the victories: "*Sewage system and dry cleaning, hygiene and washable fabrics, DDT and penicillin—the catalogue represents a series of technological victories over two great enemies: dirt and the system of untouchability, that unbrotherly contempt, to which, in the past, dirt has given rise.*" One would not expect the author of the *Brave new world*, disillusioned as he was with technical civilization, to acknowledge the achievements of technical chemistry as loudly as that. But only a Caliban's mind would reject all the tangible favours chemistry has rendered.

It took no more than a few decades for the scene to be changed. The world complains about detergents which cover the rivers with loathsome froth, detests plastic material which damages the environment, regards DDT to be the arch-enemy of the human race and even antibiotics are advised to be used with particular care.

All these might well be true. But how would the world live today without chemistry which has caused all those complaints? Is that all the thanks the chemists get?

Bibliography

Huxley A. Hyperion to a Satyr. In: Tomorrow and tomorrow and tomorrow and other essays. New York: Harper; 1956.

37

Cruel Organic Chemistry

"I will pin your ears back, Johnny, unless you eat your soup", the angry nanny kept saying. This humorous threat in the nursery might be the relic of some very unpleasant form of torture. A relic of barbarous ages our civilized world has no direct experience of—we hope. Because torturing meant the disfiguring of the body by disturbing its natural order, compelling it into some shape which was very different from the one it usually takes at rest or in unmolested motion. The aim, however, was neither killing nor maiming. The disfigured body was kept intact, and when torturing was over, it could regain its original shape, at least to a certain extent. Let the technical details be skipped now; the interested reader can find ample material together with the most modern achievements of the field.

Anyway, the looker-on must have found it easy to appreciate the cruel fate of the poor person under suspicion or being condemned. Everyone knows the way a man holds his head, where his arms and legs are, how long his waist is. The hapless sufferer had all his limbs somewhere else and in unusual states. Expressions which are related to human torture are pretty widespread in chemistry. Organic chemists speak about stretched, distorted, bent, compressed molecules or bonds and they are ready to synthetize them with predilection and great ambition. The Nobelist organic chemist Roald Hoffmann mused about that curious parlance and endeavour. He himself paid much attention to such type of compounds. Once he described a synthesis talking about the molecule in humane terms, expressing real sympathy, "The *purpose of studying unhappy molecules is not delight in their squirming under stress.* [...] *The moment one of us looked, really looked, at that poor square-planar carbon atom, he and his co-authors were thinking of a strategy to stabilize it, to give it a*

© Springer Nature Switzerland AG 2019
R. Schiller, *Between One Culture*, https://doi.org/10.1007/978-3-030-20538-6_37

chance, just a chance, for existence." Having tortured the molecule, they still sympathized with it and did their best to give it a hand. I could read compassionate sentences like these only on one other occasion, in a book by Sir Peter Medawar about guinea pigs to be experimented with.

Hoffmann talks about sympathy and help as an excuse because previously he wrote about de Sade and sadism in connection with the molecules tortured by chemists. He refers to *Pierre Laszlo* who thought "*the monstrous is a major theme of Sade. [...] Western science is underlined by its sense of wonder. Monsters are and remain an integral part of it. Wondering at monsters marks early science and carries into modern science. Such seeking for the abnormal is blatant in chemistry, in the exploration of unusual natural products and in the drive to make stressed molecules.*"

Not only the vocabulary is anthropomorphic but so are the thoughts and emotions elicited by the words. A molecule is regarded as a sensitive human being who evokes wonder, compassion and awe. Scientific ideas are in some cases expressed in terms of numerical data, equations or functions; however, in want of these, scientists have recourse to metaphors. Their *vehicle* (the direct meaning which is meant to illuminate the content called the *tenor*) often comes from human relationships. Usually, this is not realized anymore. We all understand the meaning of the sentence: "the reaction mixture attempts to oppose the change of the conditions". Certainly, the mixture is very different from the captive who attempts to escape his prison. Such metaphors, being worn out by now, have lost their emotional content.

We have seen, however, that the situation with unusual structures of molecules is different. Everyone knows how a healthy human being holds his head, where his hands and feet are; similarly, a chemist, who understands the graphic language of structural chemistry, knows what an everyday molecule looks like. Not in reality, of course, because molecules are impossible to look at, but as *Kekulé*, later *LeBel* and *van't Hoff* depicted them, and since then, with the advent of structural formulae, all chemists have tried to imagine their shapes. This is acquired knowledge, barren of any sensory impression. Methane is like *this* (Fig. 37.1), and benzene is like *that* (Fig. 37.2).

Fig. 37.1 Structure of methane

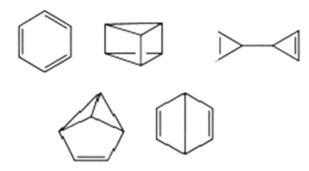

Fig. 37.2   Structure of benzene

Fig. 37.3   Benzene isomers

These shapes are regarded by chemists as healthy, whereas other, very different ones are strained, hindered, battered, unstable, hence sick, misshapen, crippled, pathetic. The ones below are isomers of benzene (Fig. 37.3). The amounts of hydrogen and carbon are the same in all of them and in the benzene seen above but, poor fellows, all of them look seriously handicapped.

In the early decades of the previous century, one could learn, thanks first to *Lewis*, later to the full understanding of quantum mechanics by *Heitler, London* and *Pauling*, that the straight lines drawn by Kekulé represent electron pairs. Even this knowledge could not discourage chemists from looking at molecules as living organisms, healthy or maimed beings.

How has it come to this? The forces acting in the solar system or in the atoms, the motion of the planets around the sun or of the electrons around the atomic nucleus have not been personified as stubbornly as that. However, gravitation or electric attraction is in one respect very different from the forces which act between the atoms in a molecule. The Sun or the Earth can attract an unlimited number of objects, as it has been witnessed by the horde of satellites. The attracting force has no definite orientation, wherever the satellite moves it is under the action of the centre. Electrostatic forces are very similar. Expressing all this in a somewhat technical manner: gravitational and electrostatic forces are not oriented and cannot be saturated.

Valence bonds are different. A carbon atom can take only four hydrogen atoms with bonds of strict directions. That is, this type of interaction is oriented and can be saturated. Well … yes, provided the molecule is healthy. We keep saying it in an anthropomorphic way because oriented and saturable interactions are unknown in the visible realm of inanimate nature. Such figures appear among men and animals only characterizing their bodily structure and mirroring their frame of mind. Molecules have neither mind nor feeling; their properties are described by the laws of quantum mechanics which, irrefutable as they are, harshly deviate from our everyday experience. But they make the molecules resemble our human nature. A carbon atom has four valences as we have four limbs and valences and limbs have their proper and natural positions. It hurts if our arm is twisted! That is the reason why the name of marquis de Sade has a place in organic chemistry.

Bibliography

Ball P. Chemistry World 2008 Jun; 36.
Hoffmann R, Hopf H. Learning from molecules in distress. Angew Chem Int Ed. 2008;47:4474–81.

38

Nobelists' Sabbath

One day before Christmas Eve 2001, BBC World TV channel organized a public debate to mark the centenary of the Nobel Prizes. The daredevil moderator was surrounded by six to eight Nobel Prize winners facing the audience made up also exclusively of Nobelists—natural scientists and economists only.

The subjects to be discussed were selected so as to arouse the layperson's interest: the aim and social use of scientific research, scientists and society, war for science, science for war. The moderator tried to urge these great men to cast a glance beyond the limits of their specialities, limits which were widened by their very own activities, asking them to tell about the effects rather than the contents of their research. He hoped to obtain a picture of the human and historical aspects of contemporary science.

The attempt was only moderately successful. A laconic Chinese-American gentleman rejected the idea, feeling unprepared to formulate any opinion beyond his own specialist field. (He might have thought the same about the other participants; this, however, remained inscrutably hidden by his Oriental smile.) "*You are much too modest*"—the moderator smiled. An American economist was clear and self-confident: "*The aim of any research is defined by the sponsor—be it the State or a private company. The scientist is commissioned to investigate a certain problem, that is his/her task.*"

Now from next to the moderator an accent, homely to my Hungarian ears, could be heard. "*It was a very poor country where I started my research work. Believe me, even with little money can one obtain nice results. I have always been controlled by my curiosity paying little heed to the eventual social use of the results I looked for.*" The speaker was *George Olah,* an alumnus of the Budapest University of Technology at *Géza Zemplén's* department. A gentleman sitting

© Springer Nature Switzerland AG 2019
R. Schiller, *Between One Culture*, https://doi.org/10.1007/978-3-030-20538-6_38

on the other side of the moderator contradicted him with a very elegant British accent. "*It was not curiosity that prompted Michelangelo to paint the frescoes of the Sistine Chapel. The scientist, before starting his work, even before being able to describe his task, knows already the question to be answered. It is not curiosity but a preliminary, tacit knowledge that controls him.*" This was the summary of the theory of *Michael Polányi*, the Hungarian-born, German and English physical chemist and philosopher. Small wonder since the remark was made by *John C. Polanyi*, his son.

Last century Europe was arguing with contemporary practical common sense.

39

Four Stamps

The portraits of four geniuses on US stamps. Great American scientists, three of them born and bred there the fourth being a native of Hungary. Two theoretical physicists, a biologist and a mathematician. The stamp collection does not enable one to learn any of these disciplines and the US Post Office certainly did not believe that, in return for 37 cents, American correspondents will be trained to become scientific geniuses. A stamp collection, however, is generally held to be an effective means of learning; it is good not only as a source of revenue but also in the education of the man of the street. As far as intellectual effort is concerned, a stamp collection is almost on a par with crossword puzzles. The display of eminent persons' portraits is not only the expression of reverence, but it also holds them up as examples. Not as far as

© Springer Nature Switzerland AG 2019
R. Schiller, *Between One Culture*, https://doi.org/10.1007/978-3-030-20538-6_39

their scientific contribution is concerned, this would obviously be absurd, but regarding their personal and human traits.

Josiah Willard Gibbs (1839–1903), together with his other important works, was one of the strict-minded founders of thermodynamics and statistical physics. He conducted the calm life of a bachelor in his sister's house. His biographer, Bumstead, finished his book about him writing: "*Unassuming in manner, genial and kindly in his intercourse with his fellow-men, never showing impatience or irritation, devoid of personal ambition of the baser sort or of the slightest desire to exalt himself, he went far toward realizing the ideal of the unselfish, Christian gentleman. In the minds of those who knew him, the greatness of his intellectual achievements will never overshadow the beauty and dignity of his life.*"

Barbara McClintock (1902–1992) had reached a venerable age when her achievements were acknowledged by a Nobel Prize. Her basic discoveries about the connections between certain hereditary properties and the structure of the chromosomes predated molecular biology. Being a woman, at the beginning of the twentieth century, she had to be exceptional in both strength and achievements to be acknowledged as a real scientist. Strong she was and what she achieved was highly appreciated by her colleagues. That was true to such an extent that her boss found her ruling personality intolerable in the research group. Nothing is known about her private life her lifelong feelings having attached her exclusively to the seeds of maize.

John von Neumann's (1903–1957) seminal ideas regarding the mathematical foundations of quantum mechanics, game theory and computer science, to name but a few of his works, are well known even to the general public. He was born in Budapest, Hungary, in a well-to-do family. The library of the spacious flat was the centre of learning and family life. Neumann read mostly history, whereas his father had a predilection for poetical works. After graduation in Budapest (mathematics) and Zürich (chemistry), the educated, always carefully dressed young man of the upper middle class moved to Berlin and soon to the United States, where he spent most of his active life. He was the contrary of an ascetic scientist, he enjoyed good food, entertained lively companies and finally came close to the circles of political power. The great physicist Eugene Wigner, a school mate of his in Budapest, said about him "*among all of us Jancsi (Neumann) was the only genius*". A fatal illness attacked him at a relatively early age. And "*John von Neumann who knew how to live a full life did not know how to die*".

Richard Feynman (1918–1988), the incredibly versatile and successful theoretical physicist, highly esteemed by his colleagues and students throughout his lifetime, a brilliant lecturer, obtained his Nobel Prize in 1965. The general opinion about him kept referring to his incredible curiosity, brilliant intellect, kind spirit and joyful mind, although he was surprisingly poorly versed in literature. He was always ready to fight, Don Quixote-like, some wind mills. He freely confessed his liberal views on sexuality and also on science: *"Physics is like sex. Sure, it may give some practical results, but that's not why we do it."* He too was killed by cancer; after two unsuccessful operations he refused any further treatment: *"I'd hate to die twice. It's so boring"*. He is said to have died with dignity.

I do not know what is taught by the four stamps.

40

A Remarkable Speech

This morning I went to the Place somewhat earlier than my usual hour. People were just coming, one by one, in the mild May weather. Still, I hoped for a sizeable audience to gather consisting of not only common gapers or horror-loving loafers but also of learned people, educated brains, well versed in scientific matters. I wished these people of merit and understanding to be witnesses of this elevating and grandiose day.

As the tumbrel was seen already in the curve of the road, I ascended the scaffold. Several of the bystanders were looking at me in bewilderment since this was no habit of the place and I was the last person who was expected to address the people from here. However, the event to come was far from being common.

"Citizens", I started in a sonorous voice, although some of my friends standing nearby could realize my cheeks to flush. "Citizens, all of us know and proclaim that liberty is nothing else but our glorious Constitution, the declaration of happiness, equality, justice and the supremacy of reason. Nothing can be estimated to be too heavy or dear, no toil, deed or sacrifice can be seen to be too burdensome if it is for the defence of our freedom against its foes at home and beyond our borders. It is not for the first time that you gather at this scaffold. Last year taught us that we, who have nailed our colours to the mast of the happiness of our nation and of all the nations, we are forced to copiously shed the blood of those who bear rancour against us. Terror has become indispensable since our Republic was established. This has been the means that made the state firm, restored obedience and discipline at home and aroused dread abroad. In order to create the liberty of mankind we had to establish the despotism of terror. This is the pure and harsh revolutionary moral."

© Springer Nature Switzerland AG 2019
R. Schiller, *Between One Culture*, https://doi.org/10.1007/978-3-030-20538-6_40

"This day, however, is different from those we have experienced so far. The man who is now nearing on the tumbrel", they were very close indeed, "was condemned to death by the Revolutionary Court due to embezzlement and trafficking with state properties. This man is the glory of not only our nation but all the human race, one of the greatest scientific minds of all ages, a learned chemist, whose achievements will always radiate a shining halo around our country and whose name will be united with our glorious revolution from today."

"Not too many years ago as his fame was known in the more educated part of the world and experts had already great interest in his results, so the rascals of the neighbouring countries became jealous of him. There was still no war between us and the German monarchs, our bayonets were not fixed against the tyrants' hearts, though his books were burnt in Berlin with great hullaba-loo. This was very similar to the way mean and dark superstition dealt with the writings suspected to be heretic in the past. This ignorant and savage act showed clearly that this scientist belongs to our country."

"There are rumours that he asked for an extension of two weeks in order to complete an experiment. This request was not granted by his judges who could thus prove their patriotic wisdom. Future ages would look upon us as playing children, we would become laughing stocks in their eyes if, amidst our work to build a new world, we had squandered the nation's time on pedantic research. All of us have to perform our duties! The judges' task was to pass a verdict, mine is to talk to you, yours to loiter here, whereas he who has been condemned must endure the punishment."

"This is a grand hour, Citizens! The fame of the Revolution's heroic deed will remain. Now a moment is worth a century. You shall never forget: to behead someone is a matter of seconds, but centuries will be unable to pro-duce a man similar to Lavoisier. This second now is ours."

I gave a sign. Sanson stepped forward, laid down the condemned and started the machine. Indeed, it took almost no time to transform the mortal scientist into the immortal of our nation.

Undeniably, this morning I returned to my lodging in a somewhat ele-vated spirit.

41

A Modest Proposal for the Lasting Improvement of the Old Age Pensioners' Quality of Life

I feel compelled to apologize to my readers, hopefully in prospect, being as bold as to leave the spheres of my learned profession and usual activity in the forthcoming pages. The anxiety of a committed citizen urges me to speak. We are being threatened by the adepts of sociology and economy with data, curves and figures. Society gets older, children are too few and people live much too long, they write, referring obviously to Europe and to areas where health systems are regarded as highly developed and birth control is exercised. In other parts of the world, influenced by the present conditions of our civilization to a lesser extent, e.g. in certain regions of Africa, the anxiety is opposite to ours: children are abundant and people's lives are much too short. We, however, have to deal with our own problems.

No seriously concerned person must look with indifference upon either the troubles which menace our economy or the weal of our aged fellow citizens, upon their much-merited, peaceful future. It seems to be a truism that the period of old age which is now being extended by years and decades, and hence the comparatively early withdrawal of an ever increasing population from the production of material goods, either aggravates the burden of the active part of the society or must reduce, undeservedly, a good many of old people to penury. Neither economic prudence nor the sense of justice of the active age groups would find admissible the continuous increase of the inactive proportion of the population, a process which is inevitable due to more cautious ways of living and the development of the medical techniques.

A solution to the problem, as recently advised by several careful thinkers, who understand the situation completely, consists in the elongation of the active period in work. Whereas the arithmetic of this proposal is simple and

© Springer Nature Switzerland AG 2019
R. Schiller, *Between One Culture*, https://doi.org/10.1007/978-3-030-20538-6_41

transparent, thus seemingly irrefutable, it conceals, tacitly, perhaps intentionally, an obvious difficulty. It assumes that long life necessarily goes hand in hand with unimpaired physical and mental abilities, stamina, creativity and love of work. It maintains that one's work at a higher age is of the same value as it was before and one might find as much joy in it as one did earlier. Sad as that is, all of us must know and experience that it is far from being true. The mechanical elongation of the active period cannot be expected to be either useful to the society or humane from the point of view of the individual. The proposition I am going to set forth below, even though in a short and rudimentary manner, is an attempt to solve the problem in a way which is useful for the society in more than one regards and at the same time it is entirely considerate to the individual.

It is difficult to understand that whereas we treat our precious partners, the farm animals, with care, sometimes even lovingly, this behaviour has not been extended to our own human community. Let us take one single example, the critical evaluation of the age data of the *blonde d'Aquitaine* cows. This is a suckler cow, its age of the first calving being 3.00 years, its full natural lifetime 6.52 years and its useful period 3.52 years. There is a clear distinction in animal husbandry between the natural lifetime and the time of slaughtering.

In order to proceed sensibly, one has to define the aim of farming. In the nineteenth century, cattle was mainly kept for the skin in Uruguay and Argentine. These animals must have been slaughtered pretty old; thus their meat was inappropriate for human consumption. In order to solve this problem, the great agricultural chemist, *Justus Liebig*, developed a method for producing an extract from the old animals' meat. This highly nutritive product, which was easy to store and to consume, became useful nourishment the industrial production of which turned out to be most profitable.

The example, clear as it is, lends itself to following. Reasonable economic thinking and natural respect to grey hair equally urge us to save both the community from excessive spending and elderly people from the pains of need and the torments of the feeling of futility. Obviously, one must avoid any rash action. When evaluating the useful span of life, one has to consider the periods of necessary education and active work, keeping in mind also the different rates of degradation of physical and mental capabilities necessary for various productive activities. Food consumption, as well, which decreases with progressing age must be taken into account.

The main aim of the present proposal is to render dignity to the ultimate period of life; those who are about to pass on should know they are still useful to society. This implies the definition of the optimum moment when industrial processing is most profitable, after our aged fellow citizens have spent a few years peacefully as well-to-do members of their communities. This is an important and difficult task to be solved by trained experts.

A critical branch of agricultural industry will soon face serious problems of raw material resources. Whereas the production of nitrate fertilizers is based on the unlimited nitrogen content of the atmosphere, the raw material of phosphate fertilizers is a product of mining. Apatite mines, however, are not inexhaustible treasure troves ready and able to feed the human race's untamed appetite without limit. In the past, animal bones were processed to produce fertilizers and also the mass graves of the battlefields were made use of to this end. Now I propose to return to this early and undeservedly forgotten method not without getting it modernized according to the conditions of present-day industrial technique and making it independent of chances and eventualities.

Standards of the physical condition of the different age groups must be elaborated and, estimating their expected values, the population of the appropriate group must be determined, with concern for the spiritual and, more importantly, the physical preparation of those who are meant for imminent industrial use. It is of little doubt that the phosphate content of the bones greatly depends on feeding in the last stage of life. Having collected all the necessary data, detailed plans for logistics and processing must be laid out in view of both industrial capacities and agricultural demand. Planning plays a pivotal role here because, on the one hand, the phosphate industry is expected to have a steady output, while, on the other hand, the age distribution of the society is optimized. The number of the inactive population must coincide with the active part's willingness to support them at a reasonable standard of living, thus expressing the esteem and reverence of the younger generation.

Although the present proposal is still somewhat sketchy and a good many of statistical and engineering details await further elaboration, its advantages are already clear. Let the main points be summarized.

1. It secures the desired age spectrum of the society for an extended period of time.
2. It solves the raw material problem of an important branch of agricultural industry.
3. It assures a reasonable living standard for people upon their retirement.
4. It ensures that those who are going to be processed by the industry can feel socially useful

If this proposal will, in some form or another, be seriously considered and at some time realized I would think it advisable to ensure that the fertilizer produced in the prospected manner be transferred to the silos in the frame of a pious and respectful ceremony.

42

The Goddess of Popular Science

On the first of January, 1801, the Sicilian astronomer, Giuseppe Piazzi, discovered a new planet, which is the first known asteroid. The planets Mercury to Saturn carry the names of Roman gods and goddesses, and Uranus was also named in the same vein when it was identified as a planet around the end of the eighteenth century. Thus, it was small wonder that the new planet was given the name of Ceres, the Roman goddess of corn, particularly because she was connected to Sicily by several ties. Two or three years later, two noted chemists, Berzelius in Sweden and Klaproth in Germany, having analysed a mineral from a Swedish mine, discovered a new element of metallic properties. Metals and planets had been associated with celestial bodies since the Antiquity: gold with the Sun, silver with the Moon, iron with Mars and mercury still preserves its relationship to Mercury in a number of languages, including English. Thus, it was quite natural that the new element was called cerium after the newly discovered asteroid. (Later it turned out that the metallic sample was a mixture of two elements, cerium and lanthanum; this fact, however, caused no trouble in naming.)

Nomenclature is habit—and perhaps something more than that. Alchemists, also connecting the properties of metals with celestial bodies, made astrology and alchemy bedfellows. The above two chemists were modern experimentalists, Dalton's contemporaries, cerium having been discovered in the decade of the birth of the Daltonian atomic theory. Nevertheless, Dalton himself denoted the elements with symbols most resembling those used by his alchemist forefathers. It is not easy to get rid of the traditions one tries to disavow.

Ceres was an ancient Italian goddess. Later, after the Roman divinities had been superseded by and transformed into the mirror image of the Greek

© Springer Nature Switzerland AG 2019
R. Schiller, *Between One Culture*, https://doi.org/10.1007/978-3-030-20538-6_42

Olympus, religious cults had become Graecized, and even diehard conservative Cato the Elder, a stubborn guardian of the ancient Roman morale and traditions, had learnt Greek; Ceres became the Latin name of the Greek Demeter. So her cult was the same as Demeter's mystery at Eleusis where, in the form of sowing and harvesting, death and resurrection was hallowed. In earlier times, however, Ceres, together with Liber and Libera, her son and daughter, was honoured in an ancient shrine near the Aventine. This was a holly place of the plebeians, whereas the patrician trinity, Jupiter, Juno and Minerva, had their seat on the Capitoline Hill. The Aventine shrine was the starting point for the people who, unhappy with the patricians, walked out to Mons Sacer; the neighbourhood of the shrine was also the site of mass assemblies and the institution of the plebs' tribune was also related to this place.

Ceres was the goddess of corn and agriculture. Ovid, however, attributes to her much more than that. As he wrote, she was "*the first to give us laws. All things are gifts from Ceres*". Living with and talking to common people, she dominated both nature and law. I believe that among all of the ancient deities, she was the one who was most closely related to the idea of nature understood by anyone.

I wonder if it was by chance or by deliberation that those who were considered by the Club of Science Journalists to be honoured with the title *The Scientist of Science Popularization of the Year* are gratified by the name of an asteroid. Anyway, the conjunction of an asteroid and popular science is an appropriate and heartening astrological symbol.

This year (2012) I was elected to be the recipient of that honour. I express my sincerest thanks to the Club and its presidium. Were I a pious Roman plebeian, I would make a thank-offering to Ceres. Thanks, however, are due to those without whom I could hardly have produced any of my popular science writings. Most of my articles appeared in the Hungarian monthly *Természet Világa* (World of Nature), established 150 years ago and continuously published ever since. The atmosphere of the journal has raised my ambition in that field. After the invitation many years ago by the then editor-in-chief, L. Dala and his co-worker, Eva Keömley, the recent decades were marked by Gyula Staar's tactfully stimulating and demanding activity; it was his formulation of tasks that elicited a good part of the texts from my computer. That is no solitary activity; it can be done only in the company of corresponding minds. It was rewarding to work under the attention of Gyula and his co-workers, Kati Kapitány and Vera Silberer. It would be fine if you kept a close eye on me in the future as well.

Bibliography

Ovid. Metamorphoses, Book V, lines 332–571 Translated by Ian Johnston. www.latinandgreek.org/_files/live/LCCS__Latin_on_the_Lawn_2012__Metamorphoses_332-571__English.pdf

A Private Letter About This Book

Professor Dr., Dr. h.c. mult. Wilhelm Ostwald
Großbothen, Landsitz Energie

Dear Professor Ostwald,

   May I take the liberty of expressing my gratitude for your kind hospitality in your house and estate called Energie, a name which reflects your solid scientific conviction. To my greatest regret I failed to meet you in person. I was more than seventy years late. Will you kindly accept my apologies? However, your dear granddaughter, a friendly and amiable elderly lady, took me round the house showing your rich library, some pieces of your laboratory equipment, entertained me with family recollections, I was even invited to lunch. Obviously, I was aware of the debt the discipline called physical chemistry owes you. Although this branch of science might still exist without your contribution (finally people discover everything), its development would have been slower, would have followed different paths, understood things in a different way and also made different errors.

   No doubt, also errors were abundant. And not just innocent ones like the ingenious but, alas, impractical construction of a mixer driven by elevating hot air. But let us recall the theory of energetics which, in your mind, superseded the naïve atomic theory of your earlier years and took you to the statement that each and any change of substance and mind or every development in the world is nothing else but variations of energy. Thus, beyond other ideas, it is of no use to assume the existence of atoms with definite mass and volume—such entities cannot be observed, anyway. Later, however, as we all know, observed they were. This happened still in your active years and it was

© Springer Nature Switzerland AG 2019
R. Schiller, *Between One Culture*, https://doi.org/10.1007/978-3-030-20538-6

the proof of your intellectual prowess and scientific integrity that you were able to revise your views. Be that as it may, refutation upon statement, contradiction upon refutation, correction upon contradiction, then… That is how even your brilliant mind worked. Is there any statement which can be regarded as final, any tenet which does not require further proof without running the risk of some falsification, any theorem the validity of which is not limited later? Do related disciplines contest? Is the best thing to do to suspend our judgement as it was suggested by some ancient thinkers? Let us be content with opinions instead of final laws?

Respect would refrain me to ask you questions like those. But how could I answer if you asked me the same? *What do I know?*

Printed in the United States
By Bookmasters